Current Topics in Microbiology and Immunology

Volume 356

Series Editors

Klaus Aktories
Medizinische Fakultät, Institut für Experimentelle und Klinische Pharmakologie
und Toxikologie, Albert-Ludwigs-Universität Freiburg, Abt. I, Albertstr. 25,
79104 Freiburg, Germany

Richard W. Compans
Influenza Pathogenesis and Immunology Research Center, Emory University,
School of Medicine, Atlanta, GA 30322, USA

Max D. Cooper
Department of Pathology and Laboratory Medicine, Georgia Research Alliance,
Emory University, 1462 Clifton Road, Atlanta, GA 30322, USA

Yuri Y. Gleba
ICON Genetics AG, Biozentrum Halle, Weinbergweg 22, 06120 Halle, Germany

Tasuku Honjo
Department of Medical Chemistry, Faculty of Medicine, Kyoto University, Sakyo-ku,
Yoshida, Kyoto, 606-8501, Japan

Hilary Koprowski
Biotechnology Foundation, Inc., 119 Sibley Avenue, Ardmore, PA 19003, USA

Bernard Malissen
Centre d'Immunologie de Marseille-Luminy, Parc Scientifique de Luminy, Case
906, 13288, Marseille Cedex 9, 13288, France

Fritz Melchers
Max Planck Institute for Infection Biology, Charitéplatz 1, 10117 Berlin, Germany

Michael B. A. Oldstone
Department of Neuropharmacology, Division of Virology, The Scripps Research
Institute, 10550 North Torrey Pines Road, La Jolla, CA 92037, USA

Peter K. Vogt
Department of Molecular and Experimental Medicine, The Scripps Research
Institute, 10550 North Torrey Pines Road, BCC-239, La Jolla, CA 92037, USA

Current Topics in Microbiology and Immunology

Previously published volumes

Further volumes can be found at www.springer.com

Vol. 326: **Reddy, Anireddy S. N.;
Golovkin, Maxim (Eds.):**
Nuclear pre-mRNA processing in plants. 2008
ISBN 978-3-540-76775-6

Vol. 327: **Manchester, Marianne;
Steinmetz Nicole F. (Eds.):**
Viruses and Nanotechnology. 2008.
ISBN 978-3-540-69376-5

Vol. 328: **van Etten, (Ed.):**
Lesser Known Large dsDNA Viruses. 2008.
ISBN 978-3-540-68617-0

Vol. 329: **Griffin, Diane E.;
Oldstone, Michael B. A. (Eds.):** Measles. 2009
ISBN 978-3-540-70522-2

Vol. 330: **Griffin, Diane E.;
Oldstone, Michael B. A. (Eds.):**
Measles. 2009.
ISBN 978-3-540-70616-8

Vol. 331: **Villiers, E. M. de (Ed.):**
TT Viruses. 2009.
ISBN 978-3-540-70917-8

Vol. 332: **Karasev A. (Ed.):**
Plant produced Microbial Vaccines. 2009.
ISBN 978-3-540-70857-5

Vol. 333: **Compans, Richard W.;
Orenstein, Walter A. (Eds.):**
Vaccines for Pandemic Influenza. 2009.
ISBN 978-3-540-92164-6

Vol. 334: **McGavern, Dorian; Dustin, Micheal (Eds.):**
Visualizing Immunity. 2009.
ISBN 978-3-540-93862-0

Vol. 335: **Levine, Beth; Yoshimori, Tamotsu;
Deretic, Vojo (Eds.):**
Autophagy in Infection and Immunity. 2009.
ISBN 978-3-642-00301-1

Vol. 336: **Kielian, Tammy (Ed.):**
Toll-like Receptors: Roles in Infection and
Neuropathology. 2009.
ISBN 978-3-642-00548-0

Vol. 337: **Sasakawa, Chihiro (Ed.):**
Molecular Mechanisms of Bacterial Infection via the Gut.
2009.
ISBN 978-3-642-01845-9

Vol. 338: **Rothman, Alan L. (Ed.):**
Dengue Virus. 2009.
ISBN 978-3-642-02214-2

Vol. 339: **Spearman, Paul; Freed, Eric O. (Eds.):**
HIV Interactions with Host Cell Proteins. 2009.
ISBN 978-3-642-02174-9

Vol. 340: **Saito, Takashi; Batista, Facundo D. (Eds.):**
Immunological Synapse. 2010.
ISBN 978-3-642-03857-0

Vol. 341: **Bruserud, Øystein (Ed.):**
The Chemokine System in Clinical
and Experimental Hematology. 2010.
ISBN 978-3-642-12638-3

Vol. 342: **Arvin, Ann M. (Ed.):**
Varicella-zoster Virus. 2010.
ISBN 978-3-642-12727-4

Vol. 343: **Johnson, John E. (Ed.):**
Cell Entry by Non-Enveloped Viruses. 2010.
ISBN 978-3-642-13331-2

Vol. 344: **Dranoff, Glenn (Ed.):**
Cancer Immunology and Immunotherapy. 2011.
ISBN 978-3-642-14135-5

Vol. 345: **Simon, M. Celeste (Ed.):**
Diverse Effects of Hypoxia on Tumor Progression. 2010.
ISBN 978-3-642-13328-2

Vol. 346: **Christian Rommel; Bart Vanhaesebroeck;
Peter K. Vogt (Ed.):**
Phosphoinositide 3-kinase in Health and Disease. 2010.
ISBN 978-3-642-13662-7

Vol. 347: **Christian Rommel; Bart Vanhaesebroeck;
Peter K. Vogt (Ed.):**
Phosphoinositide 3-kinase in Health and Disease. 2010.
ISBN 978-3-642-14815-6

Vol. 348: **Lyubomir Vassilev; David Fry (Eds.):**
Small-Molecule Inhibitors of Protein-Protein
Interactions. 2011.
ISBN 978-3-642-17082-9

Vol. 349: **Michael Kartin (Ed.):**
NF-kB in Health and Disease. 2011.
ISBN: 978-3-642-16017-2

Vol. 350: **Rafi Ahmed; Tasuku Honjo (Eds.):**
Negative Co-receptors and Ligands. 2011.
ISBN: 978-3-642-19545-7

Vol. 351: **Marcel, B. M. Teunissen (Ed.):**
Intradermal Immunization. 2011.
ISBN: 978-3-642-23689-1

Vol. 352: **Rudolf Valenta; Robert L. Coffman (Eds.):**
Vaccines against Allergies. 2011.
ISBN 978-3-642-20054-0

Cornelis Murre
Editor

Epigenetic Regulation of Lymphocyte Development

Responsible series editor: Tasuku Honjo

Springer

Cornelis Murre, Ph.D.
Department of Molecular Biology
Division of Biological Sciences
University of California, San Diego
Natural Sciences Building, Rm 5113
9500 Gilman Drive
La Jolla, CA 92093-03779
USA
e-mail: cmurre@ucsd.edu

ISSN 0070-217X
ISBN 978-3-642-24102-4   e-ISBN 978-3-642-24103-1
DOI 10.1007/978-3-642-24103-1
Springer Heidelberg Dordrecht London New York

Library of Congress Control Number: 2011942742

© Springer-Verlag Berlin Heidelberg 2012
This work is subject to copyright. All rights are reserved, whether the whole or part of the material is concerned, specifically the rights of translation, reprinting, reuse of illustrations, recitation, broadcasting, reproduction on microfilm or in any other way, and storage in data banks. Duplication of this publication or parts thereof is permitted only under the provisions of the German Copyright Law of September 9, 1965, in its current version, and permission for use must always be obtained from Springer. Violations are liable to prosecution under the German Copyright Law.
The use of general descriptive names, registered names, trademarks, etc. in this publication does not imply, even in the absence of a specific statement, that such names are exempt from the relevant protective laws and regulations and therefore free for general use.

Cover design: Deblik, Berlin

Printed on acid-free paper

Springer is part of Springer Science+Business Media (www.springer.com)

Preface

Previous observations, generated by many in the field, have provided a first glimpse into the epigenetic mechanisms that underpin lymphocyte and myeloid development. We are only now beginning to merge the multitude of observations into a common framework. At the same time it has become more difficult for the individual mind to comprehend more than a tiny focused fraction of it. The studies described in this volume serve as a starting point to familiarize one self with the multifarious differences in epigenetic designs that orchestrate the progression of developing blood cells. They also may serve as a general paradigm for the mechanisms that underpin the control of eukaryotic gene expression.

My thanks are due to the authors of this volume and Anne Clauss, Assistant Editor.

Contents

Roles of Lineage-Determining Transcription Factors in Establishing Open Chromatin: Lessons From High-Throughput Studies 1
Sven Heinz and Christopher K. Glass

B Lymphocyte Lineage Specification, Commitment and Epigenetic Control of Transcription by Early B Cell Factor 1 17
James Hagman, Julita Ramírez and Kara Lukin

Epigenetic Features that Regulate IgH Locus Recombination and Expression 39
Ramesh Subrahmanyam and Ranjan Sen

Local and Global Epigenetic Regulation of V(D)J Recombination 65
Louise S. Matheson and Anne E. Corcoran

Genetic and Epigenetic Regulation of *Tcrb* Gene Assembly 91
Michael L. Sikes and Eugene M. Oltz

T-Cell Identity and Epigenetic Memory 117
Ellen V. Rothenberg and Jingli A. Zhang

Encoding Stability Versus Flexibility: Lessons Learned From Examining Epigenetics in T Helper Cell Differentiation 145
Kenneth J. Oestreich and Amy S. Weinmann

The Epigenetic Landscape of Lineage Choice: Lessons From the Heritability of *Cd4* and *Cd8* Expression ... 165
Manolis Gialitakis, MacLean Sellars and Dan R. Littman

Index ... 189

Contributors

A. E. Corcoran Laboratory of Chromatin and Gene Expression, Babraham Research Campus, The Babraham Institute, Cambridge, CB22 3AT, UK, e-mail: anne.corcoran@bbsrc.ac.uk

M. Gialitakis Molecular Pathogenesis Program, Kimmel Center for Biology and Medicine at the Skirball Institute of Biomolecular Medicine, Howard Hughes Medical Institute, New York University School of Medicine, New York, NY 10016, USA

C. K. Glass Department of Cellular and Molecular Medicine, University of California—San Diego, La Jolla, CA 92093, USA, e-mail: ckg@ucsd.edu

J. Hagman Integrated Department of Immunology, National Jewish Health and University of Colorado School of Medicine, Denver, CO 80206, USA, e-mail: hagmanj@njhealth.org

S. Heinz Department of Cellular and Molecular Medicine, University of California—San Diego, La Jolla, CA 92093, USA, e-mail: ckg@ucsd.edu

D. R. Littman Molecular Pathogenesis Program, Kimmel Center for Biology and Medicine at the Skirball Institute of Biomolecular Medicine, Howard Hughes Medical Institute, New York University School of Medicine, New York, NY 10016, USA, e-mail: dan.littman@med.nyu.edu

K. Lukin Integrated Department of Immunology, National Jewish Health and University of Colorado School of Medicine, Denver, CO 80206, USA

L. S. Matheson Laboratory of Chromatin and Gene Expression, Babraham Research Campus, The Babraham Institute, Cambridge, CB22 3AT, UK

K. J. Oestreich Department of Immunology, University of Washington, Box 357650, 1959 NE Pacific Street, Seattle, WA 98195, USA

E. M. Oltz Department of Pathology and Immunology, Washington University School of Medicine, 660 Euclid Ave., Campus Box 8118, St. Louis, MO 63110, USA, e-mail: eoltz@pathology.wustl.edu

J. Ramírez Integrated Department of Immunology, National Jewish Health and University of Colorado School of Medicine, Denver, CO 80206, USA

E. V. Rothenberg Division of Biology 156-29, California Institute of Technology, Pasadena, CA 91125, USA, e-mail: evroth@its.caltech.edu

M. Sellars Molecular Pathogenesis Program, Kimmel Center for Biology and Medicine at the Skirball Institute of Biomolecular Medicine, Howard Hughes Medical Institute, New York University School of Medicine, New York, NY 10016, USA

R. Sen Laboratory of Molecular Biology and Immunology, Gene Regulation Section, National Institute on Aging, National Institutes of Health, 251 Bayview Boulevard, Baltimore, MD 21224, USA

M. L. Sikes Department of Microbiology, North Carolina State University, 100 Derieux Place, Campus Box 7615, Raleigh, NC 27695, USA, e-mail: mlsikes@ncsu.edu

R. Subrahmanyam Laboratory of Molecular Biology and Immunology, Gene Regulation Section, National Institute on Aging, National Institutes of Health, 251 Bayview Boulevard, Baltimore, MD 21224, USA, e-mail: rs465z@nih.gov

A. S. Weinmann Department of Immunology, University of Washington, , Box 357650, 1959 NE Pacific Street, Seattle, WA 98195, USA, e-mail: weinmann@u.washington.edu

J. A. Zhang Division of Biology 156-29, California Institute of Technology, Pasadena, CA 91125, USA

Roles of Lineage-Determining Transcription Factors in Establishing Open Chromatin: Lessons From High-Throughput Studies

Sven Heinz and Christopher K. Glass

Abstract The interpretation of the regulatory information of the genome by sequence-specific transcription factors lies at the heart of the specification of cellular identity and function. While most cells in a complex metazoan organism express hundreds of such transcription factors, the underlying mechanisms by which they ultimately achieve their functional locations within different cell types remain poorly understood. Here, we contrast various models of how cell type-specific binding patterns may arise using available evidence from ChIP-Seq experiments obtained in tractable developmental model systems, particularly the hematopoietic system. The data suggests a model whereby relatively small sets of lineage-determining transcription factors jointly compete with nucleosomes to establish their cell type-specific binding patterns. These binding sites gain histone marks indicative of active cis-regulatory elements and define a large fraction of the enhancer-like regions differentiated cell types. The formation of these regions of open chromatin enables the recruitment of secondary transcription factors that contribute additional transcription regulatory functionality required for the cell type-appropriate expression of genes with both general and specialized cellular functions.

Contents

1	Introduction	2
2	Transcription Factors Co-Localize in a Cell Type-Specific Manner	2
3	Transcription Factors Collaborate to Gain Access to Chromatinized DNA on a Genome-Wide Scale	3

S. Heinz · C. K. Glass (✉)
Department of Cellular and Molecular Medicine,
University of California, San Diego, La Jolla, CA 92093, USA
e-mail: ckg@ucsd.edu

Current Topics in Microbiology and Immunology (2012) 356: 1–15
DOI: 10.1007/82_2011_142
© Springer-Verlag Berlin Heidelberg 2011
Published Online: 10 July 2011

	3.1	Ternary Complex Formation Plays a Minor Role in Defining	
		Genome-Wide Transcription Factor Binding Patterns	3
4	Lineage-Determining Transcription Factors Prime Lineage-Specific		
	Cis-Regulatory Modules		8
	4.1	A Two-Tiered System of Transcription Factors?	8
5	Implications for Cellular Development and Reprogramming		10
	5.1	Putting it All Together	10
	5.2	Epigenomics as a Means to Identify the Factors Necessary for Cellular	
		Reprogramming?	11
References			12

1 Introduction

The genomic DNA in the cell nucleus encodes the information necessary to specify each cell in a multi-cellular organism: it predetermines the protein repertoire available to all cells of a given organism, as well as the regulatory information that orchestrates the cell type-specific expression programs necessary for development, cellular signal responses and homeostasis.

The primary genomic sequence is interpreted by sequence-specific transcription factors, and proteins that recognize and bind to sequence motifs present in the genomic sequence (Kadonaga 2004). Transcription factors act in a combinatorial fashion to recruit co-regulators, which in turn recruit RNA polymerase II and co-factors to effect transcription. Consequently, the regulatory code of the genome consists of combinations of different transcription factor motifs, also called cis-regulatory elements (CRE), which together with the inventory of transcription factors expressed in a given cell determine cell type- and developmental stage-specific transcriptomes and transcriptional programs (Davidson and Erwin 2006).

Gene deletion experiments have identified transcription factors that are essential for the generation of specific cell types (Orkin and Zon 2008). Recent technical advances have enabled analysis of the genome-wide binding patterns of these lineage-determining transcription factors. The insights from these studies highlight functional characteristics of transcription factors and how they interact with each other and with chromatin on a genome-wide level to interpret the genomic code.

2 Transcription Factors Co-Localize in a Cell Type-Specific Manner

Genome-wide studies of transcription factor localization by ChIP-Seq have revealed a surprising variability in transcription factor binding patterns (cistromes) of a given factor in different cell types or at different stages of development (Cao et al. 2010; Heinz et al. 2010; Jakobsen et al. 2007; Krum et al. 2008; Lefterova et al. 2010; Lin et al. 2010; Lupien et al. 2008; Odom et al. 2004; Palii et al. 2010; Sandmann et al. 2006, 2007; Verzi et al. 2010). Surprisingly, the majority of the cistrome differences are restricted to promoter-distal inter- and intragenic sites. At the same time, different factors in the same cell type tend to co-localize to these sites on a genome-wide scale, which correlates with the expression (Boyer et al. 2005; Chen et al. 2008; Heinz et al. 2010; Lefterova et al. 2010; Li et al. 2008; Lin et al. 2010; MacArthur et al. 2009; Sullivan et al. 2010; Verzi et al. 2010; Wilson et al. 2010).

To date, the following factors have been observed to co-localize in a cell type-specific fashion in hematopoietic cells: RUNX1 and ETS1 in Jurkat T-ALL cells as a model for CD4$^+$ T cells (Hollenhorst et al. 2009), TAL1 with RUNX1 and ETS1 in Jurkat cells as well as TAL1 with GATA1 in erythroblasts (Palii et al. 2010), STAT3 and IRF-4 in IL-21-treated CD4$^+$ T cells (Kwon et al. 2009), SCL, LYL1, GATA2, LMO2, ERG, FLI-1 and RUNX1 in the hematopoietic progenitor line HPC-7 (Wilson et al. 2010), E2A, EBF1 and FOXO1 in Rag1$^{-/-}$ pro-B cells (Lin et al. 2010), PU.1 and C/EBPα/β in macrophages (Heinz et al. 2010; Lefterova et al. 2010), as well as PU.1 and OCT-2 in splenic B cells (Heinz et al. 2010), PPARγ and PU.1 and C/EBPβ in macrophages (Lefterova et al. 2010) and SRF and PU.1 in macrophages (Sullivan et al. 2010).

3 Transcription Factors Collaborate to Gain Access to Chromatinized DNA on a Genome-Wide Scale

3.1 Ternary Complex Formation Plays a Minor Role in Defining Genome-Wide Transcription Factor Binding Patterns

The mechanisms underlying this cell type-specific and differential genomic targeting of transcription factors remain poorly understood (Farnham 2009). Several hypotheses have been put forward to explain both observations. A large majority of the co-bound regions harbor the consensus motifs for the respective factors, and these motifs do not differ between cell types or differentiation stages. Therefore, both a tethering mechanism by which one factor binds to the region via another that makes contact with the DNA and differential targeting of factors to varying motifs due to, for example, post-translational modification of the binding specificity of the factors can be excluded.

Perhaps the most commonly proposed explanation is the assumption of protein–protein interactions, which stabilize transcription factor-DNA interactions and thus contribute to combinatorial targeting of transcription factors to different locations in the genome (Lodish et al. 2007). This would account for both the co-localization of transcription factors in a given cell type, and the differences in cistromes if one of the interaction partners is not expressed or replaced by another one with different specificity in a different cell type.

Cooperative DNA binding and targeting can indeed be observed when analyzing motifs for factors known to participate in ternary complexes: for example, PU.1-IRF half-site composite motifs with fixed distance between the respective motifs are highly enriched in PU.1-bound regions in both macrophages and B cells (Fig. 1b). IRF8 and IRF4 expressed are known to bind DNA only as ternary complexes with PU.1 in these cells types (Eisenbeis et al. 1993; Pongubala et al. 1992). Similarly, close analysis of the distance relationships between the enriched sequence motifs in various transcription factor ChIP-Seq experiments reveals that the motifs for some co-localizing factors frequently occur at a fixed distance from each other (e.g. composites of ETS1:RUNX1 (Hollenhorst et al. 2009), C/EBP:AP-1 (Heinz et al. 2010) and E2A:PU1 (Lin et al. 2010)), and several of these have been confirmed by co-immunoprecipitation and electrophoretic mobility shift assays.

Ternary complex formation imposes a distance requirement on the sequence motifs for the interacting transcription factors, which can be either fixed or within a very limited sequence distance range, depending on the anatomy of the involved proteins (Ogata et al. 2003). However, for many of the co-bound factors, and even for the above factors that can bind as ternary complexes, the majority of the distance distributions for their co-occurring motifs is bell-shaped with maximum distances of up to 150 bp (corresponding to 50 nm, 10 times the diameter of a globular protein of 40 kDa) (Fig. 1b and (Heinz et al. 2010; Lin et al. 2010)). This distance range by far exceeds the combined size of the proteins themselves, such that direct protein–protein interactions above ~ 20 bp would have to involve looping of the intervening DNA. While looping has been observed for e.g. c-myb and C/EBPβ on the Mim-1 promoter at a motif distance of 82 bp (Ogata et al. 2003), the stiffness of intervening DNA stretches shorter than that would prohibit direct protein–protein interactions. Additional bending by, for example, HMG domain proteins (Love et al. 1995) could contribute to tighter bends in the intervening DNA and allow closer interaction distances, but motifs for these bending factors should also be co-enriched at these sites. For both adjacent and looping-mediated protein–protein interaction, the helical nature and torsional rigidity of DNA would require phasing of the motifs for the cooperating factors. Together, this would lead to a tri-modal motif co-occurrence frequency distribution exhibiting motif phasing with a 10.4-base period as depicted in Fig. 1c, and possibly additional motifs to be enriched at co-bound sites, which is contradicted by the bell-shaped motif distributions and the enriched motif sets observed.

In summary, while ternary complex formation accounts for a small fraction of the observed targeting of transcription factors to the genome, it does not explain the majority of the observations.

3.1.1 A Case for Chromatin

Other suggested explanations for the observed differences in genomic targeting of transcription factors involve the epigenome: in its natural state in a living cell, the genomic DNA is packaged into chromatin, and the presence of nucleosomes

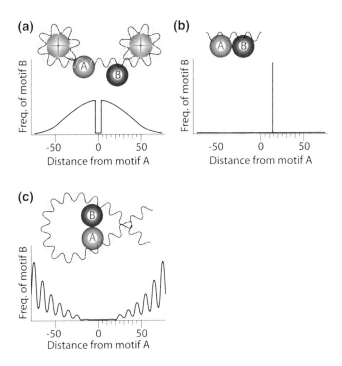

Fig. 1 Different modes of transcription factor co-binding give rise to characteristic motif frequency distributions. Transcription factors ("A" and "B") are depicted as globular proteins of ∼40 kDa/5 nm diameter. Illustrations are approximately to scale. **a** Collaborative binding of transcription factors in competition with nucleosomes. The schematic drawing illustrates binding of two transcription factors to their cognate sites, thereby stably displacing nucleosomes (tetrapartite circles circled by DNA) to flanking regions. *Bottom* Representation of the predicted bell-shaped cumulative motif frequency distribution of one of the two factors when centering genomic regions co-bound by both factors in this fashion on the motif of the other factor. This is the most commonly observed type of motif distribution at co-bound genomic sites in ChIP-Seq experiments. **b** Ternary complex formation without looping. Direct protein–protein-DNA interactions without long-range bending of DNA impose strict motif distance requirements. **c** Ternary complex formation with looping. Long-range bending of the intervening DNA stabilizes ternary protein–protein–DNA interactions. Binding to motifs that are spaced too closely and would require severe bending is prevented by the inherent stiffness of DNA. The torsional rigidity of DNA below 400 bp length prevents protein–protein interactions at motifs that are not facing each other, leading to phasing of the motif frequency pattern

restricts transcription factor access to the genomic information (Beato and Eisfeld 1997).

One hypothesis proposes that "active" histone modifications, particularly histone H3 lysine 4 mono- and dimethylation (H3K4me1/2), marks of active chromatin (Heintzman et al. 2007) may direct binding of transcription factors (Lupien et al. 2008). This is in line with previous observations that transcription factors tend to bind to regions marked by the aforementioned H3K4me1 (Robertson et al. 2008; Heintzman et al. 2009). However, the majority of transcription factors is not expected to directly interact with modified histone tails, and additionally, this leaves unanswered the question how the histone modifications are targeted to different regions in different cell types in a sequence-specific fashion in the first place.

3.1.2 Pioneer Factors Collaborate to Access the Chromatinized Genome

Alternatively, "pioneer factors" with the ability to bind to nucleosomal DNA and to displace nucleosomes may generate open chromatin, and provide DNA access to secondary transcription factors in a cell type-specific manner (Cirillo et al. 2002; Ghisletti et al. 2010). This hypothesis places pioneer factors at the top of a hierarchy of transcription factors, in line with their essential functions in lineage specification and organogenesis (Zaret et al. 2008; Natoli 2010; Smale 2010). In this context, it is important to note that many pioneer factors are expressed in multiple cell types, where they exhibit cell-specific cistromes (e.g. FOXA1 (Lupien et al. 2008), MyoD (Cao et al. 2010), PU.1 (Heinz et al. 2010) or C/EBPβ (Lefterova et al. 2010). True to the definition of a pioneer factor, these factors would be expected to bind to their genomic binding sites irrespective of cell type and presence of other transcription factors, and to localize to the same genomic locations in all cell types, which is not in agreement with the observed ChIP-Seq and ChIP-chip results.

Intriguingly, de novo analysis of the motifs enriched at sites of cell type-specific binding of both "pioneer factors" and factors not deemed "pioneers", shows that transcription factors prefer to bind to genomic locations that contain both their cognate motifs as well as motifs for lineage-determining "master regulators", or pioneer factors (Benner et al. unpublished). The motif distance distribution of co-binding factors is bell-shaped, and, as described above, extends to about 150 bp, or approximately one nucleosome. This suggests a mechanism for genome-wide targeting whereby binding of multiple transcription factors to adjacent motifs leads to nucleosome displacement and stable transcription factor occupancy (Adams and Workman 1995; Boyes and Felsenfeld 1996; Miller and Widom 2003). This hypothesis accounts for the characteristics of the observed differential co-localization discussed above: It is independent of protein–protein interactions (Adams and Workman 1995; Vashee et al. 1998), and transcription factor targeting to the genome depends on the complement of transcription factors and their concentrations in a given cell. We will use the term "collaborative" binding to distinguish this mode of joint binding to neighboring sites in the absence of direct protein–protein interactions from cooperative binding due to direct protein–protein interactions.

3.1.3 Genome-Wide Data Supports a Collaborative Mechanism of Transcription Factor Binding Patterning

A collaborative mechanism for cell type-specific targeting of transcription factors is supported by several lines of evidence: Analysis of the cistromes of PU.1 in B cell progenitors arrested at different stages due to ablation of either

E2A, EBF1 or Rag1 reveals differential genome-wide targeting of PU.1 to PU.1 motifs in the vicinity of motifs for B cell-specific transcription factors that are expressed and essential for progression of B cell development to the respective stage (Heinz et al. 2010). Similarly, the E2A cistrome in EBF-deficient pre-pro-B cells is depleted of E2A binding sites in the vicinity of EBF motifs relative to the E2A cistrome in EBF-expressing $Rag1^{-/-}$ pro-B cells (Lin et al. 2010). Additional evidence comes from retroviral reconstitution experiments in transcription factor-deficient progenitors: C/EBPβ in myeloid and PU.1 in B lymphoid progenitors lacking PU.1 or E2A, respectively, gain large numbers of additional binding sites upon retroviral complementation with the missing factors (Heinz et al. 2010). This occurs both upon short-term transduction of E2A knock-out cells with E2A, or activation of a stably expressed PU.1-ER fusion protein with tamoxifen in the case of PU.1 knock-out cells. The latter is already present in the nucleus at low concentration in the un-induced state, and tamoxifen treatment raises the nuclear concentration of the PU.1-ER fusion protein by only ninefold (S. Heinz, unpublished observation). Nonetheless, this leads to a marked increase in the number of C/EBPβ-bound sites genome-wide in the vicinity of PU.1 sites that gain PU.1-ER occupancy, highlighting the role of different concentrations of collaborating factors in combinatorially defining their genome-wide cistromes. The motifs for the respective factors at binding sites gained upon reconstitution exhibit no specific distance requirements, indicating absence of direct protein–protein interactions and collaborative combinatorial targeting of PU.1 and associated factors to nucleosomal targets sites.

3.1.4 Transcription Factor Concentrations Define the Joint Binding Pattern

The notion of collaboration between transcription factors easily follows the importance of transcription factor motif clustering in the absence of strong sequence conservation, which has been analyzed in detail for the even-skipped enhancer in different fly species (Hare et al. 2008), and an importance for low-affinity binding sites in gene regulatory networks (Segal et al. 2008). This model also predicts a role for neighboring nucleosome positioning sequences in determining the threshold binding concentrations for a given motif cluster (Mirny 2010). For example, high versus low concentrations of PU.1 are required for macrophage and B cell differentiation, respectively (DeKoter and Singh 2000). If we assume targeting of PU.1 to be independent of other factors, and only to depend on the affinity of PU.1 for a given motif, the lower PU.1 concentration in B cells versus macrophages should lead to a smaller cistrome in B cells, which should be a subset of the one observed in macrophages. This notion is contradicted by the genome-wide data: while it is true that the PU.1 cistrome in B cells comprises less sites, it is not a mere subset of the macrophage cistrome (Heinz et al. 2010), and similar cistromic differences have been observed for other factors in disparate cell types (see above). In contrast, when assuming that PU.1 localization depends on the total combined concentrations of PU.1 and other factors that bind to neighboring sites to exceed the threshold energy for stably displacing nucleosomes from the motifs, then varying the concentrations of PU.1 alone would change the observable PU.1 binding pattern according to the cellular concentrations of the other transcription factors.

4 Lineage-Determining Transcription Factors Prime Lineage-Specific Cis-Regulatory Modules

4.1 A Two-Tiered System of Transcription Factors?

Motif analysis of the genome-wide binding sites non-lineage-determining factors such as BCL-6 and NF-κB (Barish et al. 2010), LXRβ (Heinz et al. 2010) or SRF (Sullivan et al. 2010) in macrophages and STAT3 in IL-21-treated CD4$^+$ T cells (Kwon et al. 2009) reveals that binding of non-lineage-determining factors frequently occurs in the vicinity of the lineage-determining factors that are essential for the generation of a given cell type. Likewise, loss of the lineage-determining factor leads to loss of binding of non-lineage-determining factors at a large fraction in their sites.

For example, sites of IL-21-induced STAT3 binding in CD4$^+$ T cells are highly co-enriched for the motif for IRF-4 (Kwon et al. 2009), and conditional deletion of *Irf4* is accompanied by a dramatic loss of STAT3 binding at 85% of the sites bound in wild-type CD4$^+$ T cells. However, in stark contrast to the effects of lineage-determining factors, which appear to affect the cistromes of all other transcription factors, loss or drastic increase in nuclear concentration of a non-lineage-determining factor leaves the cistrome of lineage-determining factors unperturbed: activating STAT3 in CD4$^+$ T cells with IL-21 does not change the IRF-4 cistrome. Similarly, for the lineage-determining factor PU.1, ablation of the genes encoding for the non-lineage-determining factors *Bcl6*, *Lxrα/β* or *Srf* in macrophages does not affect the PU.1 cistrome (Barish et al. 2010; Heinz et al. 2010; Sullivan et al. 2010), while PU.1 is required to define a large fraction of their binding patterns.

This indicates a functional disparity between these two types of factors in terms of their ability to define the landscape of accessible chromatin; whereas lineage-determining factors together appear to influence the locations bound by all transcription factors in a given cell genome-wide, non-lineage-determining factors seem unable to influence binding-site selection by lineage-restricted transcription factors.

4.1.1 Lineage-Determining Factors Induce Histone H3K4 Mono-and Di-Methylation Surrounding Their Binding Sites

Given the essential roles that promoters play in initiating and regulating transcription, the finding that the majority of genome-wide differences in transcription factor binding in different cell types do not occur at transcription start sites (TSS)

Roles of Lineage-Determining Transcription Factors

was somewhat surprising. In fact, the majority of cell type-specific binding of transcription factors occurs at promoter-distal sites (Cao et al. 2010; Heintzman et al. 2009; Heinz et al. 2010; Lin et al. 2010).

This is in line with the important roles of distal regulatory elements in regulating cell type-specific gene expression patterns: enhancers, repressors and locus-control regions add an additional regulatory layer (Bulger and Groudine 2009), leading to the stable, and high-magnitude differences in gene expression and regulatory responses observed between disparate cell types in higher eukaryotes.

The majority of promoter-distal transcription factor binding sites is flanked by a histone modification pattern comprising high levels of histone H3 mono- and dimethylated on lysine 4 (H3K4me1^{hi}) and low to undetectable levels of trimethylated histone H3K4 (H3K4me3^{lo}) (Lupien et al. 2008; Robertson et al. 2008; Ghisletti et al. 2010; Heinz et al. 2010; Lin et al. 2010), a histone modification pattern first described as a mark of putative enhancers (Heintzman et al. 2007). The H3K4me3^{lo}/H3K4me1^{hi} signature distinguishes these distal sites from promoters, which are characterized by a distinct H3K4me3^{hi}/H3K4me1^{lo} pattern (Heintzman et al. 2007).

Recent studies have revealed that the deposition of this mark is targeted by transcription factor binding: Data for PU.1 in macrophages and myeloid progenitors and E2A in pre-pro-B cells indicate that dynamic nucleosome displacement and stable binding of lineage-determining factors precedes the deposition of the aforementioned H3K4me1 mark to distal CREs (Heinz et al. 2010; Lin et al. 2010), and loss of PU.1 in macrophages leads to concomitant loss of H3K4me1 in the vicinity of previously PU.1-bound sites (Ghisletti et al. 2010). Similar observations have recently been made in retinoic acid-induced differentiation of mouse pluripotent P19 cells to neural cells for FOXA1 (Sérandour et al. 2011), a prototypic pioneer factor with critical roles in the formation of multiple tissues during embryonic development (Friedman and Kaestner 2006). In contrast to the effects of lineage-determinants, gain or loss of non-lineage-determining transcription factors such as LXRα/β (Heinz et al. 2010), STAT1 (Robertson et al. 2008) or SRF (Sullivan et al. 2010) does not significantly alter the H3K4me1 pattern around their binding sites. Correspondingly, the motif sets most highly enriched in H3K4me1^{hi}/H3K4me3^{lo} promoter-distal regions in a wide range of cell types comprise those of the essential tissue-specific transcriptional regulators that are necessary for the generation of the respective cell type, while motifs for non-lineage-determining transcription factors are not significantly enriched in these same regions (Heinz et al. 2010).

Thus, targeting of H3K4 mono- and dimethylation to promoter-distal sites appears to be another characteristic feature of lineage-determining transcription factors.

4.1.2 Mechanistic Notes

We would like to speculate as to how and why methyl marks are being deposited in the vicinity of promoter-distal binding sites of lineage-determining transcription factors. Conceivably, lineage-determining factors could directly interact with and recruit methyltransferases to their binding sites, which would lead to H3K4 methylation flanking all binding sites of a given factor. This is unlikely, given that $\sim 25\%$ of the PU.1 sites bound de novo in an inducible model of macrophage differentiation do not exhibit histone H3K4 methylation on flanking nucleosomes (Heinz et al. 2010). A hint at a possible mechanism comes from the H3K4 methylation profiles seen around promoter-distal sites, which are similar to the H3K4 methylation profiles several kilobases downstream of promoters (H3K4me1 > m2 > me3), and reverse those observed directly at promoters (H3K4me3 > me2 > me1). Therefore, we suggest that promoter-distal H3K4 mono- and dimethylation around transcription factor binding sites is deposited by promoter-proximal MLL/COMPASS complexes that are brought into the vicinity of the promoter-distal sites by the transcription-enhancing actions of the bound transcription factors.

4.1.3 Signal-Dependent Transcription Factors Act on Primed CREs

In contrast to most lineage-determining factors, which exhibit cell type-restricted expression patterns, many of the non-lineage-determining factors that do not significantly affect the open-chromatin landscape are signal-dependent transcription factors.

Rather than change the shape of the chromatin landscape, these secondary factors appear to "read" it by binding to the exposed fraction of their cognate motifs (Ghisletti et al. 2010; Heinz et al. 2010). Consequently, the cistromes of, for example, STAT3 are drastically different in IL-21-treated CD4$^+$ T cells (Kwon et al. 2009) and embryonic stem cells (Chen et al. 2008), where in each case, STAT3 binding predominantly occurs in the vicinity of binding sites for cell type-defining transcription factors. Secondary factors recruit additional co-activators such as p300 (Ghisletti et al. 2010; Barish et al. 2010), CBP (Kim et al. 2010) or co-repressors e.g. HDACs (and likely entire co-repressor complexes) (Barish et al. 2010), to shape the transcriptome according to the cell type-specific chromatin landscape that has been set up by the combinatorial action of cell type-specifying regulators.

5 Implications for Cellular Development and Reprogramming

5.1 Putting it All Together

Single-locus studies have provided detailed insight into how promoter-distal CREs gain competence, which involves step-wise opening and recruitment of transcription factors (e.g. see Decker et al. 2009; Hoogenkamp et al. 2009). Together with the insights from genome-wide studies discussed above, this paints a picture whereby successive activation of lineage-determining transcription factors enables their combinatorial recruitment to chromatin. Their continued action leads to remodeling and histone modification, or more generally, to epigenomic events that together define the developmental trajectory of open and accessible chromatin. The cell type-specific complement of open chromatin defines the accessible genomic binding sites for second-tier factors, which translate signal responses into cell type-specific transcriptional outcomes. The defining features that distinguish primary and secondary factors remain to be determined, but they conceivably include nuclear concentrations and the ability to interact with the chromatin remodelers expressed in a given cell type/stage or activation state. Thus, whether a given factor functions as primary "master regulator" or a secondary factor likely depends on the cellular context.

Given that if lineage-determining transcription factors define regions of open chromatin, they might also function to interpret the genomic code by shaping the three-dimensional structure of the genome (Fullwood et al. 2009; Lieberman-Aiden et al. 2009; Natoli 2010), by serving as adaptors for or facilitating binding of structural proteins such as mediator, cohesin or CTCF, which mediate promoter-enhancer interactions (Kagey et al. 2010; Schmidt et al. 2010), and may be involved in defining chromatin topological changes observed, for example, at antigen receptor loci during B cell and T cell development (Jhunjhunwala et al. 2009).

5.2 Epigenomics as a Means to Identify the Factors Necessary for Cellular Reprogramming?

The notion that binding of lineage-determining transcription factors leads to the conjugation of methyl groups to histone H3K4 on neighboring nucleosomes is corroborated by the results of de novo analysis of motifs associated with H3K4me1$^+$ regions: in all cell types examined, the motifs most highly enriched are consensus motifs for the essential transcription factors driving the development of a given lineage (Heinz et al. 2010). Namely in macrophages, the most highly enriched H3K4me1-enriched motifs are consensus motifs for ETS, C/EBP, AP-1 and RUNX. Conversely, in H3K4me1$^+$ regions in embryonic stem cells, the predominantly enriched sequence motifs are for KLF, SOX and OCT factors as well as ESRRB. Members of these respective protein families are sufficient to reprogram various cell types into induced macrophages (Xie et al. 2004; Laiosa et al. 2006; Feng et al. 2008; Bussmann et al. 2009) and pluripotent stem cells (Takahashi and Yamanaka 2006; Feng et al. 2009), which suggests that assessing the H3K4me1-associated motif pattern and identifying the overlap with the set of expressed transcription factors may represent a facile way to pinpoint the transcription factors necessary for cellular reprogramming into a given cell type.

5.2.1 Outlook

In conclusion, high-throughput sequencing-based methods not only continue to produce novel insight into how differentiation processes both shape and are being shaped by the epigenome, but they also allow a renewed look at the very basic mechanisms that gene regulatory networks operate on top of, serving as a genome-wide complement to detailed singe-locus studies.

References

Adams CC, Workman JL (1995) Binding of disparate transcriptional activators to nucleosomal DNA is inherently cooperative. Mol Cell Biol 15(3):1405–1421

Barish GD, Yu RT, Karunasiri M, Ocampo CB, Dixon J, Benner C, Dent AL, Tangirala RK, Evans RM (2010) Bcl-6 and NF-kappaB cistromes mediate opposing regulation of the innate immune response. Genes Dev 24(24):2760–2765

Beato M, Eisfeld K (1997) Transcription factor access to chromatin. Nucleic Acids Res 25(18): 3559–3563

Boyer LA, Lee TI, Cole MF, Johnstone SE, Levine SS, Zucker JP, Guenther MG, Kumar RM, Murray HL, Jenner RG, Gifford DK, Melton DA, Jaenisch R, Young RA (2005) Core transcriptional regulatory circuitry in human embryonic stem cells. Cell 122(6):947–956. doi: S0092-8674(05)00825-1[pii]

Boyes J, Felsenfeld G (1996) Tissue-specific factors additively increase the probability of the all-or-none formation of a hypersensitive site. EMBO J 15(10):2496–2507

Bulger M, Groudine M (2009) Enhancers: the abundance and function of regulatory sequences beyond promoters. Dev Biol 339(2):250–257

Bussmann LH, Schubert A, Vu Manh TP, De Andres L, Desbordes SC, Parra M, Zimmermann T, Rapino F, Rodriguez-Ubreva J, Ballestar E, Graf T (2009) A robust and highly efficient immune cell reprogramming system. Cell Stem Cell 5(5):554–566. doi:S1934-5909(09)00515-3[pii]

Cao Y, Yao Z, Sarkar D, Lawrence M, Sanchez GJ, Parker MH, MacQuarrie KL, Davison J, Morgan MT, Ruzzo WL, Gentleman RC, Tapscott SJ (2010) Genome-wide MyoD binding in skeletal muscle cells: a potential for broad cellular reprogramming. Dev Cell 18(4):662–674

Chen X, Xu H, Yuan P, Fang F, Huss M, Vega VB, Wong E, Orlov YL, Zhang W, Jiang J, Loh YH, Yeo HC, Yeo ZX, Narang V, Govindarajan KR, Leong B, Shahab A, Ruan Y, Bourque G, Sung WK, Clarke ND, Wei CL, Ng HH (2008) Integration of external signaling pathways with the core transcriptional network in embryonic stem cells. Cell 133(6): 1106–1117. doi:S0092-8674(08)00617-X[pii]

Cirillo LA, Lin FR, Cuesta I, Friedman D, Jarnik M, Zaret KS (2002) Opening of compacted chromatin by early developmental transcription factors HNF3 (FoxA) and GATA-4. Mol Cell 9(2):279–289

Davidson EH, Erwin DH (2006) Gene regulatory networks and the evolution of animal body plans. Science 311(5762):796–800

Decker T, Pasca di Magliano M, McManus S, Sun Q, Bonifer C, Tagoh H, Busslinger M (2009) Stepwise activation of enhancer and promoter regions of the B cell commitment gene Pax5 in early lymphopoiesis. Immunity 30(4):508–520. doi:S1074-7613(09)00136-8[pii]

DeKoter RP, Singh H (2000) Regulation of B lymphocyte and macrophage development by graded expression of PU.1. Science 288(5470):1439–1441. doi:8531[pii]

Eisenbeis CF, Singh H, Storb U (1993) PU.1 is a component of a multiprotein complex which binds an essential site in the murine immunoglobulin lambda 2–4 enhancer. Mol Cell Biol 13(10):6452–6461

Farnham PJ (2009) Insights from genomic profiling of transcription factors. Nat Rev Genet 10(9):605–616

Feng R, Desbordes SC, Xie H, Tillo ES, Pixley F, Stanley ER, Graf T (2008) PU.1 and C/EBPalpha/beta convert fibroblasts into macrophage-like cells. Proc Natl Acad Sci USA 105(16):6057–6062. doi:0711961105[pii]

Feng B, Jiang J, Kraus P, Ng JH, Heng JC, Chan YS, Yaw LP, Zhang W, Loh YH, Han J, Vega VB, Cacheux-Rataboul V, Lim B, Lufkin T, Ng HH (2009) Reprogramming of fibroblasts into induced pluripotent stem cells with orphan nuclear receptor Esrrb. Nat Cell Biol 11(2): 197–203. doi:ncb1827[pii]

Friedman JR, Kaestner KH (2006) The Foxa family of transcription factors in development and metabolism. Cell Mol Life Sci 63(19–20):2317–2328

Fullwood MJ, Liu MH, Pan YF, Liu J, Xu H, Mohamed YB, Orlov YL, Velkov S, Ho A, Mei PH, Chew EG, Huang PY, Welboren WJ, Han Y, Ooi HS, Ariyaratne PN, Vega VB, Luo Y, Tan PY, Choy PY, Wansa KD, Zhao B, Lim KS, Leow SC, Yow JS, Joseph R, Li H, Desai KV, Thomsen JS, Lee YK, Karuturi RK, Herve T, Bourque G, Stunnenberg HG, Ruan X, Cacheux-Rataboul V, Sung WK, Liu ET, Wei CL, Cheung E, Ruan Y (2009) An oestrogen-receptor-alpha-bound human chromatin interactome. Nature 462(7269):58–64

Ghisletti S, Barozzi I, Mietton F, Polletti S, De Santa F, Venturini E, Gregory L, Lonie L, Chew A, Wei CL, Ragoussis J, Natoli G (2010) Identification and characterization of enhancers controlling the inflammatory gene expression program in macrophages. Immunity 32(3):317–328

Hare EE, Peterson BK, Iyer VN, Meier R, Eisen MB (2008) Sepsid even-skipped enhancers are functionally conserved in Drosophila despite lack of sequence conservation. PLoS Genet 4(6):e1000106

Heintzman ND, Stuart RK, Hon G, Fu Y, Ching CW, Hawkins RD, Barrera LO, Van Calcar S, Qu C, Ching KA, Wang W, Weng Z, Green RD, Crawford GE, Ren B (2007) Distinct and predictive chromatin signatures of transcriptional promoters and enhancers in the human genome. Nat Genet 39(3):311–318. doi:ng1966[pii]

Heintzman ND, Hon GC, Hawkins RD, Kheradpour P, Stark A, Harp LF, Ye Z, Lee LK, Stuart RK, Ching CW, Ching KA, Antosiewicz-Bourget JE, Liu H, Zhang X, Green RD, Lobanenkov VV, Stewart R, Thomson JA, Crawford GE, Kellis M, Ren B (2009) Histone modifications at human enhancers reflect global cell-type-specific gene expression. Nature 459(7243):108–112. doi:nature07829[pii]

Heinz S, Benner C, Spann N, Bertolino E, Lin YC, Laslo P, Cheng JX, Murre C, Singh H, Glass CK (2010) Simple combinations of lineage-determining transcription factors prime cis-regulatory elements required for macrophage and B cell identities. Mol Cell 38(4):576–589

Hollenhorst PC, Chandler KJ, Poulsen RL, Johnson WE, Speck NA, Graves BJ (2009) DNA specificity determinants associate with distinct transcription factor functions. PLoS Genet 5(12):e1000778 (Epub 2009 Dec 18)

Hoogenkamp M, Lichtinger M, Krysinska H, Lancrin C, Clarke D, Williamson A, Mazzarella L, Ingram R, Jorgensen H, Fisher A, Tenen DG, Kouskoff V, Lacaud G, Bonifer C (2009) Early chromatin unfolding by RUNX1: a molecular explanation for differential requirements during specification versus maintenance of the hematopoietic gene expression program. Blood 114(2):299–309. doi:blood-2008-11-191890[pii]

Jakobsen JS, Braun M, Astorga J, Gustafson EH, Sandmann T, Karzynski M, Carlsson P, Furlong EE (2007) Temporal ChIP-on-chip reveals Biniou as a universal regulator of the visceral muscle transcriptional network. Genes Dev 21(19):2448–2460. doi:21/19/2448[pii]

Jhunjhunwala S, van Zelm MC, Peak MM, Murre C (2009) Chromatin architecture and the generation of antigen receptor diversity. Cell 138(3):435–448

Kadonaga JT (2004) Regulation of RNA polymerase II transcription by sequence-specific DNA binding factors. Cell 116(2):247–257

Kagey MH, Newman JJ, Bilodeau S, Zhan Y, Orlando DA, van Berkum NL, Ebmeier CC, Goossens J, Rahl PB, Levine SS, Taatjes DJ, Dekker J, Young RA (2010) Mediator and cohesin connect gene expression and chromatin architecture. Nature 467(7314):430–435

Kim TK, Hemberg M, Gray JM, Costa AM, Bear DM, Wu J, Harmin DA, Laptewicz M, Barbara-Haley K, Kuersten S, Markenscoff-Papadimitriou E, Kuhl D, Bito H, Worley PF, Kreiman G, Greenberg ME (2010) Widespread transcription at neuronal activity-regulated enhancers. Nature 465(7295):182–187

Krum SA, Miranda-Carboni GA, Lupien M, Eeckhoute J, Carroll JS, Brown M (2008) Unique ERalpha cistromes control cell type-specific gene regulation. Mol Endocrinol 22(11): 2393–2406. doi:me.2008-0100[pii]10.1210/me.2008-0100

Kwon H, Thierry-Mieg D, Thierry-Mieg J, Kim HP, Oh J, Tunyaplin C, Carotta S, Donovan CE, Goldman ML, Tailor P, Ozato K, Levy DE, Nutt SL, Calame K, Leonard WJ (2009) Analysis of interleukin-21-induced Prdm1 gene regulation reveals functional cooperation of STAT3 and IRF4 transcription factors. Immunity 31(6):941–952

Laiosa CV, Stadtfeld M, Xie H, de Andres-Aguayo L, Graf T (2006) Reprogramming of committed T cell progenitors to macrophages and dendritic cells by C/EBP alpha and PU.1 transcription factors. Immunity 25(5):731–744. doi:S1074-7613(06)00475-4[pii]10.1016/j.immuni.2006.09.011

Lefterova MI, Steger DJ, Zhuo D, Qatanani M, Mullican SE, Tuteja G, Manduchi E, Grant GR, Lazar MA (2010) Cell-specific determinants of peroxisome proliferator-activated receptor gamma function in adipocytes and macrophages. Mol Cell Biol 30(9):2078–2089 (Epub 2010 Feb 22)

Li XY, MacArthur S, Bourgon R, Nix D, Pollard DA, Iyer VN, Hechmer A, Simirenko L, Stapleton M, Luengo Hendriks CL, Chu HC, Ogawa N, Inwood W, Sementchenko V, Beaton A, Weiszmann R, Celniker SE, Knowles DW, Gingeras T, Speed TP, Eisen MB, Biggin MD (2008) Transcription factors bind thousands of active and inactive regions in the Drosophila blastoderm. PLoS Biol 6(2):e27. doi:07-PLBI-RA-2717[pii]10.1371/journal.pbio.0060027

Lieberman-Aiden E, van Berkum NL, Williams L, Imakaev M, Ragoczy T, Telling A, Amit I, Lajoie BR, Sabo PJ, Dorschner MO, Sandstrom R, Bernstein B, Bender MA, Groudine M, Gnirke A, Stamatoyannopoulos J, Mirny LA, Lander ES, Dekker J (2009) Comprehensive mapping of long-range interactions reveals folding principles of the human genome. Science 326(5950):289–293

Lin YC, Jhunjhunwala S, Benner C, Heinz S, Welinder E, Mansson R, Sigvardsson M, Hagman J, Espinoza CA, Dutkowski J, Ideker T, Glass CK, Murre C (2010) A global network of transcription factors, involving E2A, EBF1 and Foxo1, that orchestrates B cell fate. Nat Immunol 11(7):635–643 (Epub 2010 Jun 13)

Lodish H, Berk A, Kaiser CA, Krieger M, Scott MP, Bretscher A, Ploegh H, Matsudaira P (2007) Molecular cell biology. 6th edn. W.H.Freeman & Co, New York

Love JJ, Li X, Case DA, Giese K, Grosschedl R, Wright PE (1995) Structural basis for DNA bending by the architectural transcription factor LEF-1. Nature 376(6543):791–795

Lupien M, Eeckhoute J, Meyer CA, Wang Q, Zhang Y, Li W, Carroll JS, Liu XS, Brown M (2008) FoxA1 translates epigenetic signatures into enhancer-driven lineage-specific transcription. Cell 132(6):958–970. doi:S0092-8674(08)00118-9[pii]10.1016/j.cell.2008.01.018

MacArthur S, Li XY, Li J, Brown JB, Chu HC, Zeng L, Grondona BP, Hechmer A, Simirenko L, Keranen SV, Knowles DW, Stapleton M, Bickel P, Biggin MD, Eisen MB (2009) Developmental roles of 21 Drosophila transcription factors are determined by quantitative differences in binding to an overlapping set of thousands of genomic regions. Genome Biol 10(7):R80. doi:gb-2009-10-7-r80[pii]10.1186/gb-2009-10-7-r80

Miller JA, Widom J (2003) Collaborative competition mechanism for gene activation in vivo. Mol Cell Biol 23(5):1623–1632

Mirny LA (2010) Nucleosome-mediated cooperativity between transcription factors. Proc Natl Acad Sci 107(52):22534–22539

Natoli G (2010) Maintaining cell identity through global control of genomic organization. Immunity 33(1):12–24

Odom DT, Zizlsperger N, Gordon DB, Bell GW, Rinaldi NJ, Murray HL, Volkert TL, Schreiber J, Rolfe PA, Gifford DK, Fraenkel E, Bell GI, Young RA (2004) Control of pancreas and liver gene expression by HNF transcription factors. Science 303(5662):1378–1381. doi:10.1126/science.1089769303/5662/1378[pii]

Ogata K, Sato K, Tahirov TH (2003) Eukaryotic transcriptional regulatory complexes: cooperativity from near and afar. Curr Opin Struct Biol 13(1):40–48

Orkin SH, Zon LI (2008) Hematopoiesis: an evolving paradigm for stem cell biology. Cell 132(4):631–644

Palii CG, Perez-Iratxeta C, Yao Z, Cao Y, Dai F, Davidson J, Atkins H, Allan D, Dilworth FJ, Gentleman R, Tapscott SJ, Brand M (2010) Differential genomic targeting of the transcription factor TAL1 in alternate haematopoietic lineages. EMBO J 2010:21

Pongubala JM, Nagulapalli S, Klemsz MJ, McKercher SR, Maki RA, Atchison ML (1992) PU.1 recruits a second nuclear factor to a site important for immunoglobulin kappa 3′ enhancer activity. Mol Cell Biol 12(1):368–378

Robertson AG, Bilenky M, Tam A, Zhao Y, Zeng T, Thiessen N, Cezard T, Fejes AP, Wederell ED, Cullum R, Euskirchen G, Krzywinski M, Birol I, Snyder M, Hoodless PA, Hirst M, Marra MA, Jones SJ (2008) Genome-wide relationship between histone H3 lysine 4 mono- and tri-methylation and transcription factor binding. Genome Res 18(12):1906–1917. doi:gr.078519.108[pii]

Sandmann T, Jensen LJ, Jakobsen JS, Karzynski MM, Eichenlaub MP, Bork P, Furlong EE (2006) A temporal map of transcription factor activity: mef2 directly regulates target genes at all stages of muscle development. Dev Cell 10(6):797–807. doi:S1534-5807(06)00170-5[pii]

Sandmann T, Girardot C, Brehme M, Tongprasit W, Stolc V, Furlong EE (2007) A core transcriptional network for early mesoderm development in *Drosophila melanogaster*. Genes Dev 21(4):436–449. doi:21/4/436[pii]

Schmidt D, Schwalie PC, Ross-Innes CS, Hurtado A, Brown GD, Carroll JS, Flicek P, Odom DT (2010) A CTCF-independent role for cohesin in tissue-specific transcription. Genome Res 20(5):578–588

Segal E, Raveh-Sadka T, Schroeder M, Unnerstall U, Gaul U (2008) Predicting expression patterns from regulatory sequence in Drosophila segmentation. Nature 451(7178):535–540

Sérandour AA, Avner S, Percevault F, Demay F, Bizot M, Lucchetti-Miganeh C, Barloy-Hubler F, Brown M, Lupien M, Metivier R, Salbert G, Eeckhoute J (2011) Epigenetic switch involved in activation of pioneer factor FOXA1-dependent enhancers. Genome Res 21(4):555–565 (Epub 2011 Jan 13)

Smale ST (2010) Pioneer factors in embryonic stem cells and differentiation. Curr Opin Gen Dev 20(5):519–526 (Epub 2010 Jul 16)

Sullivan AL, Benner C, Heinz S, Huang W, Xie L, Miano JM, Glass CK (2010) SRF utilizes distinct promoter and enhancer-based mechanisms to regulate cytoskeletal gene expression in macrophages. Mol Cell Biol 31(4):861–875 (Epub 2010 Dec 6)

Takahashi K, Yamanaka S (2006) Induction of pluripotent stem cells from mouse embryonic and adult fibroblast cultures by defined factors. Cell 126(4):663–676. doi:S0092-8674(06)00976-7[pii]

Vashee S, Melcher K, Ding WV, Johnston SA, Kodadek T (1998) Evidence for two modes of cooperative DNA binding in vivo that do not involve direct protein–protein interactions. Curr Biol 8(8):452–458

Verzi MP, Shin H, He HH, Sulahian R, Meyer CA, Montgomery RK, Fleet JC, Brown M, Liu XS, Shivdasani RA (2010) Differentiation-specific histone modifications reveal dynamic chromatin interactions and partners for the intestinal transcription factor CDX2. Dev Cell 19(5):713–726. doi:10.1016/j.devcel.2010.10.006

Wilson NK, Foster SD, Wang X, Knezevic K, Schutte J, Kaimakis P, Chilarska PM, Kinston S, Ouwehand WH, Dzierzak E, Pimanda JE, de Bruijn MF, Göttgens B (2010) Combinatorial transcriptional control in blood stem/progenitor cells: genome-wide analysis of ten major transcriptional regulators. Cell Stem Cell 7(4):532–544

Xie H, Ye M, Feng R, Graf T (2004) Stepwise reprogramming of B cells into macrophages. Cell 117(5):663–676. doi:S0092867404004192[pii]

Zaret KS, Watts J, Xu J, Wandzioch E, Smale ST, Sekiya T (2008) Pioneer factors, genetic competence, and inductive signaling: programming liver and pancreas progenitors from the endoderm. Cold Spring Harb Symp Quant Biol 73:119–126

B Lymphocyte Lineage Specification, Commitment and Epigenetic Control of Transcription by Early B Cell Factor 1

James Hagman, Julita Ramírez and Kara Lukin

Abstract Early B cell factor 1 (EBF1) is a transcription factor that is critical for both B lymphopoiesis and B cell function. EBF1 is a requisite component of the B lymphocyte transcriptional network and is essential for B lineage specification. Recent studies revealed roles for EBF1 in B cell commitment. EBF1 binds its target genes via a DNA-binding domain including a unique 'zinc knuckle', which mediates a novel mode of DNA recognition. Chromatin immunoprecipitation of EBF1 in pro-B cells defined hundreds of new, as well as previously identified, target genes. Notably, expression of the pre-B cell receptor (pre-BCR), BCR and PI3K/Akt/mTOR signaling pathways is controlled by EBF1. In this review, we highlight these current developments and explore how EBF1 functions as a tissue-specific regulator of chromatin structure at B cell-specific genes.

Contents

1 Prologue	18
2 Introduction...	18
3 Early B Cell Factor 1: Protein Structure and Function	19
3.1 Early Studies of EBF1	19
3.2 X-ray Crystallographic Analysis of EBF1 Structure....................................	21
3.3 Structures of Other Domains in EBF1	22
4 Control of B Lymphopoiesis Requires a Network of Proteins Including EBF1	22
4.1 EBF1 and the Basis of B Lineage Fate Decisions.....................................	24
4.2 Effects of Changes in *Ebf1* Gene Dosage on B Cell Development......................	26

J. Hagman (✉) · J. Ramírez · K. Lukin
Integrated Department of Immunology,
National Jewish Health and University of Colorado School of Medicine,
Denver, CO 80206, USA
e-mail: hagmanj@njhealth.org

Current Topics in Microbiology and Immunology (2012) 356: 17–38
DOI: 10.1007/82_2011_139
© Springer-Verlag Berlin Heidelberg 2011
Published Online: 7 July 2011

	4.3 EBF1 and B Cell Lineage Commitment	27
5	Identification of EBF1 Target Genes Using Genome-Wide Analysis	28
	5.1 Binding of EBF1 to a Vast Array of Sites in Pro-B Cells	28
	5.2 Co-occupancy of Genes by EBF1 and E2A	29
	5.3 Epigenetic Regulation of Genes by EBF1	30
	5.4 Multiple Mechanisms Activate *Cd79a* Transcription in Early B Cells	31
6	Conclusions	33
References		34

1 Prologue

Twenty years ago, Early B cell factor 1 (EBF1; first identified as EBF) was detected as a novel DNA-binding activity specific for the *Cd79a* (*mb-1*) promoter (Hagman et al. 1991; Feldhaus et al. 1992). This activity was restricted to nuclear extracts from B cells. The discovery of EBF1 generated much excitement because it was identified as a potential regulator of B lymphocyte lineage specification and commitment. These roles of EBF1 were supported eventually by extensive data. A series of seminal observations concerning EBF1's structure, functions and mechanisms of action in B lymphopoiesis were made in the past 3 years. Here, we review the recent literature concerning EBF1 and its roles in the production of B cells, the antibody-producing arm of the immune system.

2 Introduction

B lymphocytes produce antibodies in response to antigenic challenges. In the bone marrow, these cells are generated from multi-potent progenitors (MPPs) that have the ability to become a variety of hematopoietic cells. A key event during B cell differentiation is the expression of Early B cell Factor 1 (EBF1), which drives the specification of B lineage cells in concert with other DNA-binding proteins including E2A and Pax5. How EBF1 accomplishes these functions has been a mystery. Until very recently, even the structure of EBF1 and its mode of DNA binding were unknown. Structural determinations have resolved many of these issues and are discussed here in detail.

The ability of EBF1 to direct the differentiation of uncommitted progenitors is a function of two of its intrinsic properties: EBF1 (1) activates transcription of B cell specific genes including *Pax5*, which encodes a B lineage commitment factor (O'Riordan and Grosschedl 1999) and (2) enforces commitment by repressing the expression of drivers of alternative lineages (such as C/EBPα and Id2) (Pongubala et al. 2008; Thal et al. 2009). Prior to lineage commitment of hematopoietic progenitor cells, EBF1 directs expression of the B cell program and represses other programs. Recent data suggest that B lineage specification and commitment are each affected by the dosage of EBF1.

In the past, biochemical methods identified a small number of potential gene targets of EBF1. More recently, chromatin immunoprecipitation (ChIP) was used to isolate DNA occupied by EBF1 in pro-B cells (Lin et al. 2010; Treiber et al. 2010b). These studies enabled the characterization of sequences bound by EBF1 in vivo. The experiments revealed an unexpectedly high number of promoter, enhancer and intergenic sites bound by EBF1. These DNA sequences were often clustered with binding sites of other regulators within the B lineage network (E2A, Runx1 and FOXO1). These studies also revealed epigenetic signatures of activated and repressed genes in pro-B cells. An important conclusion of these reports is that EBF1 is essential for initiating epigenetic changes in target genes during early B cell differentiation. However, these activities require prior modifications of chromatin and/or other factors, which may be responsible for lineage priming that precedes B lymphopoiesis. The nature of these signals and their origins is unknown and is a focus of speculation below.

An important distinction between EBF1 and other transcription factors is its ability to activate the B cell program by epigenetic remodeling of chromatin in early B cell progenitors. In this regard, EBF1 may interact directly with co-activators and SWI/SNF chromatin remodeling complexes. At the early B cell-specific *Cd79a* promoter, binding of EBF1 results in increased chromatin accessibility and decreased DNA methylation (Gao et al. 2009). Recent genomic analyses of histone modifications suggest a model of progressive gene activation predicated upon modifications initiated prior to the expression of EBF1, as well as those that are critically dependent on EBF1 itself.

3 Early B Cell Factor 1: Protein Structure and Function

3.1 Early Studies of EBF1

EBF1 (also known as EBF, O/E-1 and COE1) is a member of the EBF family of transcription factors. Early studies detected EBF1 binding to a functionally important palindromic site within the early B cell-specific *Cd79a* promoter (Hagman et al. 1991; Feldhaus et al. 1992). The *Cd79a* promoter drives expression of Ig-α, a transmembrane protein that is essential for display of the pre-B cell receptor (pre-BCR) and the BCR on the B cell plasma membrane as well as for signaling functions (Hombach et al. 1990; Campbell et al. 1991; Gold et al. 1991). Biochemical studies of EBF1 revealed that it assembles stable homodimers in the absence of DNA (Travis et al. 1993). Cloning and sequencing of cDNAs encoding EBF1 revealed its novel protein sequence (Fig. 1a) (Hagman et al. 1993; Wang and Reed, 1993). The major isoform of EBF1 is 591 amino acids, which includes a $\sim$215 residue DNA-binding domain (DBD) and an atypical helical region comprising the helix-loop-helix-loop-helix (HLHLH) domain.

Biochemical and mutational studies confirmed the function of the DBD. The amino acid sequence alignment of the DBD with those of other known DBDs

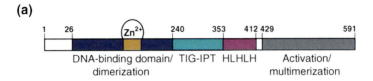
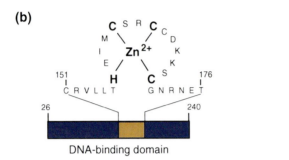
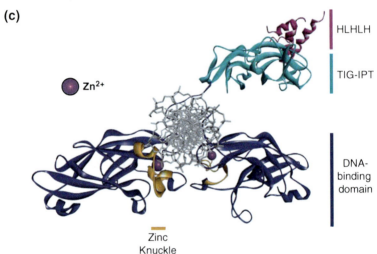

Fig. 1 *The structure of Early B cell factor 1 (EBF1).* The domains of EBF are labeled and colored consistently throughout. **a** A schematic representation of the domains in $EBF1_{1-591}$ includes the DNA-binding domain's (DBD) unique zinc knuckle (*gold*). The DBD (*blue*), TIG/IPT (*teal*) and HLHLH (*magenta*) domains all participate in EBF dimerization. The carboxyl terminus includes the Ser/Thr/Pro-rich activation domain. The amino acids demarcating each domain are numbered. **b** An expanded view of the zinc knuckle highlights the histidine and three cystine residues that coordinate the zinc ion required for DNA binding. **c** The structure of an $EBF1_{26-422}$ dimer bound to DNA (*grey*). The perspective is parallel to the helical axis of the DNA molecule. The visible portion of the HLHLH domain, the TIG/IPT domain, the DBD domains and the zinc knuckle motifs are indicated. The zinc ions are depicted as purple spheres. The structure was generated using PDB file ID 3MLP (Treiber et al. 2010a) and was modeled using Discovery Studio Visualizer 3.0, Accelrys Inc., San Diego, CA

detected only very limited sequence identity (14% with the p65 subunit of NF-κB; (Siponen et al. 2010). EBF1 homodimers bind efficiently to inverted repeat DNA sequences consisting of two half-sites that are separated by a two base pair spacer. In vitro measurements suggested that the optimal nucleotide target sequence of EBF1 is 5'-ATTCCCNNGGGAAT-3' (Hagman et al. 1995). Although it lacks consensus zinc fingers, the ability of EBF1 to bind DNA is dependent on its incorporation of zinc ions. Mutagenesis studies suggested that the metal ion is coordinated by a single histidine and three cysteines within a fourteen residue motif. The motif is termed the zinc knuckle and is required for DNA binding (Fig. 1b) (Hagman et al. 1995; Fields et al. 2008).

Dimerization of EBF1 is essential for its function. The HLHLH domain of EBF1 was predicted to include three putative α-helical motifs (H1, H2A and H2B) similar to those identified in basic-HLH family proteins (such as MyoD1); however, these proteins only have H1 and H2 (Hagman et al. 1993). Homodimerization of EBF1 requires contributions of the DBD, HLHLH and the Transcription factor Immuno-globulin (Ig)/Ig Plexin-like fold in Transcription factors (TIG/IPT) domain between the DBD and HLHLH (Hagman et al. 1993; Hagman et al. 1995; Aravind and Koonin 1999). The carboxy-terminal region of EBF1 is enriched with serine, threonine and proline residues and potently activates transcription when appended to a heterolo-gous DBD (Hagman et al. 1995). Together, these studies helped delineate functionally important sequences in EBF1, but they did little to reveal how the protein folds, binds DNA or functions in vivo.

3.2 X-ray Crystallographic Analysis of EBF1 Structure

Recent structural determinations confirmed that EBF1 is the founding member of a distinct family of DNA-binding proteins. Structural characterization of the EBF1 DBD revealed a 'pseudoimmunoglobulin' fold similar to those of Rel family proteins (Siponen et al. 2010; Treiber et al. 2010a). The overall fold includes a core consisting of an anti-parallel β-barrel that contains nine β-strands arranged in two interacting sheets. An amino-terminal α-helix packs against the bottom of this structure. A series of loops that extend from the Rel-like core are among the distinct features of the EBF1 DBD. Protruding from the rest of the DBD, the zinc knuckle coor-dinates zinc ions using three short α-helices within the His-X_3-Cys-X_2-Cys-X_5-Cys motif. This configuration, which is one of the smallest independently-folding protein domains, is different from other types of zinc fingers (reviewed in Klug 2010).

EBF1 (residues 26–240 or 26–422) crystallized as a dimer bound to an optimal palindromic DNA sequence (Fig. 1c) (Treiber et al. 2010a). The structure revealed much concerning DNA recognition by EBF1 and of other closely related family members (e.g. EBF3). The complex has several novel features. Three distinct motifs within each subunit of the paired EBF1 homodimer make contacts with the major and minor grooves of the palindromic site. A highly unusual feature is the recognition of bases within both half-sites by each EBF1 monomer. Residues within an extended loop between the β-strands of the DBD's amino-terminus and a carboxyl-terminal loop recognize the invariant bases within the major groove of one half-site. The zinc knuckle makes contacts with the minor groove of the other half-site. Thus, contacts made by each monomer within the homodimer assemble a symmetric clamp that extends across both half-sites of the palindrome. This configuration explains the requirement for a two base pair spacer between the two half-sites recognized by EBF1 (Travis et al. 1993). Mutagenesis of contact residues confirmed their importance for DNA binding (Fields et al. 2008; Siponen et al. 2010; Treiber et al. 2010a).

The DNA sequence used for crystallization with EBF1 (including 5'-ATT-CCCATGGGAAT-3') and the majority of EBF1 binding sites identified using ChIP and bioinformatics are highly palindromic. In contrast, the EBF1 binding site of the *Cd79a* promoters (5'-AGACTCAAGGGAAT-3') is less symmetric. Thus, the clamping mode of DNA binding may be critical for the ability of EBF1 to activate promoters with less than optimal binding sites.

3.3 Structures of Other Domains in EBF1

Regions of EBF1 involved in homodimerization were crystallized both as individual domains (Siponen et al. 2010) and in the context of dimers of residues 26–422 bound to DNA (Treiber et al. 2010a). Folding of the TIG/IPT domain in an Ig-like structure similar to Rel family members was confirmed. Some similarities were noted between the TIG/IPT domain and human calmodulin-binding transcription activator 1 (CAMTA1), a member of a family of proteins that includes highly conserved DBDs (CG-1) (Finkler et al. 2007). Packing of the TIG/IPT domain of EBF3 in crystals was used to model interactions between interfaces of the homologous domains in EBF1, suggesting that they contribute to the formation of multimers in solution (Siponen et al. 2010). The HLHLH domain was defined only weakly in the context of the DNA-bound EBF1 complex (26–422), with only one helix-loop-helix motif visible in the complex (Treiber et al. 2010a). However, comparisons of EBF1 with related b-HLH proteins (i.e., homodimers of E47; (Ellenberger et al. 1994) approximates how the HLHLH may mediate homodimerization. More studies are needed to reveal contributions of these domains to the DNA binding and function of EBF1.

4 Control of B Lymphopoiesis Requires a Network of Proteins Including EBF1

Hematopoietic stem cells (HSCs) are the source of all major blood cell lineages. B cells develop from HSCs following a series of progressively restricted rounds of differentiation in the bone marrow (Fig. 2) (reviewed in Kondo 2010). Expansion

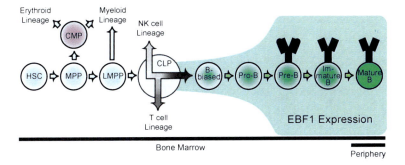

Fig. 2 *EBF1 expression in the context of hematopoiesis.* Beginning with HSCs (hematopoietic stem cells), progressive steps of cell-fate restriction result in the designated cell types and lineages. MPPs (multi-potent progenitors) give rise to CMPs (common-myeloid progenitors), which possess myeloid and lymphoid potential. Additionally, MPPs yield LMPPs (lymphoid-primed multi-potent progenitors), which possess myeloid, lymphoid and NK cell potential. Common-lymphoid progenitors (CLPs), can be trisected into **a** $Rag1^{lo}\lambda5^-$ cells which are confined to the NK lineage, **b** $Rag1^{hi}\ \lambda5^-$ cells which can develop into NK cells or T lymphocytes and **c** $Rag1^{hi}\ \lambda5^+$ cells which retain NK, T and B cell potential. Pre-B cells express the pre-B cell receptor (pre-BCR), which is composed of Ig heavy chains and the surrogate light chain proteins VpreB1 and $\lambda5$. Immature and mature B cells express mature BCRs comprised of Ig heavy and light chains. The relative levels of EBF1 expression in CLPs through mature B cells are represented by the size of the *green shaded block*. Developmental steps that occur in the bone marrow versus the periphery are indicated

of short-term HSCs is followed by the production of MPPs, which have the ability to produce all hematopoietic lineages, but which lack self-renewing capacity. Among the descendants of these cells, lymphoid-primed multi-potent progenitors (LMPPs) are the earliest lineage cells to express receptors for Interleukin-7 (IL-7) and maintain the ability to generate both myeloid and lymphoid cells (Adolfsson et al. 2005). LMPPs differentiate into common lymphoid progenitors (CLPs), which were first identified as $Lin^-IL-7R\alpha^+c-Kit^{Lo}Sca1^+$ cells in bone marrow (Kondo 1997). It is now understood that CLPs constitute heterogeneous populations that produce B, T or natural killer (NK) cells (reviewed in Ichii et al. 2010).

Generation of LMPPs requires transcription factors including PU.1 (*Sfpi1*), Gfi-1 (*Gfi1*), Ikaros (*Ikzf1*) and E2A (multiple proteins expressed by the *Tcfe2a* gene) (Reviewed in Ramirez et al. 2010), which regulate cells' decisions to attain myeloid versus lymphoid fates. The concentration, or dosage, of regulatory factors helps determine the priming of cell differentiation (lineage priming) and subsequent fate decisions. For example, the control of alternative myeloid or B lymphoid fates of MPPs is a function of high or low concentrations of PU.1, respectively (Dekoter and Singh 2000). Levels of PU.1 expression are specified by the zinc finger protein Gfi-1, which promotes the development of B cells by repressing *Sfpi1*, the gene that encodes PU.1 (Spooner et al. 2009). Ikaros directs hematopoietic progenitors by promoting B cell-specific gene expression and *Ig heavy chain* gene rearrangements (Reynaud et al. 2008). Lineage priming is one of the mechanisms by which Ikaros directs cell fates toward lymphopoiesis

(Ng et al. 2009). This activity of Ikaros is essential for B and T cell development. These processes are promoted directly by Ikaros' binding to gene targets and indirectly via its restriction of other transcriptional programs including those of early progenitors and non-lymphoid lineages. Lineage priming by Ikaros may be a consequence of its establishment of epigenetic modifications in chromatin at genes that will be expressed in a tissue-specific manner following lineage commitment.

E2A also received attention recently as an essential primer of B lymphopoiesis. Although expressed in both B and T cells, early studies showed that ablation of E2A expression in mice resulted in the complete loss of B lineage cells (Bain et al. 1994; Zhuang et al. 1994). The generation of hematopoietic cells including CLPs, and subsequently, lymphoid cells, was dependent on the expression of E2A proteins in HSCs (Dias et al. 2008; Semerad et al. 2009). E2A proteins also promoted lymphoid development and suppressed myeloid fates in a dose-dependent fashion. E2A-dependent lineage priming of lymphocyte-specific genes may promote lymphoid fates. One of the proteins encoded by *Tcfe2a*, E12, drives B cell development in *Tcfe2a*$^{-/-}$ MPPs (Bhalla et al. 2008). A consequence of E12 expression is the restoration of EBF1 expression, which is lost in the absence of E2A. These activities of E2A proteins, together with co-occupancy of regulatory modules with EBF1, suggest the importance of functional collaborations between the two factors in B lymphopoiesis.

4.1 EBF1 and the Basis of B Lineage Fate Decisions

Many studies have implicated EBF1 as a primary determinant of B cell lineage specification. The lack of EBF1 in *Ebf1* knockout mice results in complete developmental arrest at a CLP-like stage of development (Lin and Grosschedl 1995). Having lost a key regulator of B cell development, it is not surprising that these mice exhibit: (1) loss of B cell-specific gene expression including key proteins required for differentiation (*Cd79a*, *B29/Cd79b*, *Vpreb1*, *Igll1*(λ5) and *Rag1*) and (2) a complete absence of V(D)J recombination, which is necessary to assemble functional *Ig* genes. In contrast, enforced expression of EBF1 in murine HSCs drives B cell development at the expense of other hematopoietic lineages (Zhang et al. 2003). Moreover, in non-lymphoid cells, EBF1 can activate at least part of the B cell program in the absence of other upstream regulators (Kee and Murre 1998; Romanow et al. 2000; Goebel et al. 2001; Medina et al. 2004; Pongubala et al. 2008).

The central role of EBF1 in the B cell-specific network of transcription factors was confirmed by Medina et al. (2004). The differentiation of PU.1-deficient progenitor cells is arrested completely at the MPP stage. In these mutant cells, enforced expression of EBF1 rescued B cell development, resulting in pro-B cells that expressed key B cell-specific genes (e.g. the genes that fail to be expressed in EBF1-deficient mice) and activated V(D)J recombination. While *Pax5* genes were turned on by EBF1, enforced expression of Pax5 alone did not rescue B cell development similarly.

During normal hematopoiesis, EBF1 is first expressed at low levels in CLPs (Dias et al. 2005; Roessler et al. 2007). Comparisons of CLPs derived from wild-type versus EBF1-deficient mice revealed important roles of EBF1 in B cell specification and commitment (Zandi et al. 2008). In the absence of EBF1, populations of CLPs failed to express transcripts of key genes required for functional B cells, including *Cd79a*, *B29/Cd79b* and *Igll1*. Additionally, EBF1 was required for the activation of transcription factor genes *Pax5*, *Pou2af1* (OCA-B/BOB-1/OBF1) and *Foxo1*, which play important roles at later stages of B cell development (Kim et al. 1996; Schubart et al. 1996; Nutt et al. 1999; Hess et al. 2001; Dengler et al. 2008; Herzog et al. 2008; Srinivasan et al. 2009). EBF1-deficient CLPs also failed to initiate *Ig* heavy chain gene *D–J* recombination. Together, these data provide evidence for lineage priming of B cell-specific genes by EBF1 in CLPs.

More recently, CLPs were sub-divided into three populations that possess different lineage potentials related to their expression of EBF1 (Månsson et al. 2010). CLPs that expressed low levels of both EBF1 and a GFP reporter of *Rag1* expression gave rise to NK, B and T cells. Increased *Rag1*/GFP expression correlated with increased *Ebf1* expression and restriction of lineage potential to B and T lymphocytes. The mice used in these studies also possessed a *Igll1* promoter:human CD25 transgene, which served as a reporter for $\lambda5$ surrogate light chain expression. Single-cell multiplex PCR analysis confirmed that $Rag1^{hi}\lambda5^+$ CLPs correlated with the highest frequency of *Ebf1* expression, expressed *Pax5* transcripts and were restricted solely to the B lineage (similar to Hardy fraction A cells). The authors concluded that EBF1 plays a crucial role in CLPs to direct B cell lineage commitment (together with Pax5 and Ikaros; Nutt et al. 1999; Reynaud et al. 2008).

How EBF1 is activated in a subset of B cell progenitors is an open question. Expression of EBF1 is likely driven by E2A proteins. Regulation of EBF1 may also involve FOXO1 because FOXO1 binding sites have been found in regulatory regions across the *Ebf1* locus (Kee and Murre 1998; Smith et al. 2002; Lin et al. 2010). *Ebf1* transcription may be regulated by IL-7/STAT5 signaling as well (Purohit et al. 2003; Dias et al. 2005; Kikuchi et al. 2005; Roessler et al. 2007). However, a recent report demonstrated that EBF1 is expressed in STAT5-deficient pro-B cells that have been rescued by Bcl-2 (Malin et al. 2010). Activation of *Ebf1* genes may involve a stochastic mechanism in which activation of a single allele is a consequence of low levels of activating signals in a subset of CLPs. Additionally, EBF1 may promote its own expression by repressing Id2 and Id3 expression, which in turn suppress E2A activity (Pongubala et al. 2008; Thal et al. 2009). Once activated by EBF1, expression of Pax5 increases EBF1 expression via a positive feedback loop (Roessler et al. 2007). Moreover, EBF1 has been reported to upregulate its own expression (Smith et al. 2002). The data are consistent with a cascade of factors that drive cells toward increasing production of EBF1 by direct positive regulation (Rothenberg 2007), which promotes the B cell fate while limiting other lineage choices.

4.2 Effects of Changes in Ebf1 Gene Dosage on B Cell Development

As described above, transcription of *Ebf1* genes begins in CLPs, where the factor biases cells toward B lymphopoiesis. As cells progress from CLPs to functional B cells, they pass through multiple intermediate stages (i.e. Hardy fractions) as identified by the sequential expression of cell surface markers and V(D)J recombination (reviewed in Hardy et al. 2007) (Fig. 2). Progression through the various stages is characterized by the expression of increasing amounts of EBF1, which is generated by complex regulation of the *Ebf1* gene's two promoters (Roessler et al. 2007). In a process that is linked with upregulation of Pax5, levels of *Ebf1* transcripts increase significantly between 'B-biased' cells (Hardy fraction A) and pro-B cells (fraction B). *Ebf1* transcripts are upregulated (>five-fold) in the transition of pro-B cells to pre-B cells (fractions C' and D; Roessler et al. 2007; H. Lei and J.H., unpublished data). The basis of the upregulation in fractions C' and D is unknown. However, expression of different concentrations of EBF1 as development progresses suggests that its activities are, in part, dosage-dependent.

Dosage-dependent effects of EBF1 have been confirmed in heterozygous knock-out mice with single functional *Ebf1* alleles. Numbers of B cell progenitors and mature B cells were reduced by half in fetal livers, bone marrow and spleens of *Ebf1*$^{+/-}$ mice (Lin and Grosschedl 1995; O'Riordan and Grosschedl 1999; Lukin et al. 2010). This effect was compounded by combining the *Ebf1*$^{+/-}$ genotype with haplo-insufficiency of *Tcfe2a* or *Runx1* genes, which encode transcription factors that bind DNA cooperatively with EBF1. EBF1:E2A (E47) complexes assemble on promoters including *Igll1* (Sigvardsson et al. 1997). Combined haplo-insufficient mice with single functional *Ebf1* and *Tcfe2a* alleles (*Ebf1*$^{+/-}$*Tcfe2a*$^{+/-}$) displayed defective development of pro-B cells together with reduced expression of B cell-specific transcripts including *Pax5*, *Rag1*, *Rag2*, *Vpreb1*, *Igll1* and *Cd79a*. These studies also included the first observations that EBF1 directly regulates *Pax5* promoter transcription. Together, these studies indicated the importance of collaborative interactions between EBF1 and E2A proteins.

EBF1 also binds DNA cooperatively with the Runx1 transcription factor, which was first described at the *Cd79a* promoter (Maier et al. 2004). Bone marrow B cell development exhibited a striking compound phenotype in *Ebf1*$^{+/-}$*Runx1*$^{+/-}$, or *ER*^{het} mice (Lukin et al. 2010). Most effects were apparent at the pro-B-pre-B boundary and at subsequent stages. This disruption occurs later than effects observed in *Ebf1*$^{+/-}$*Tcfe2a*$^{+/-}$ mice. Reduced levels of EBF1 alone decreased the frequency of *Igλ light chain* gene rearrangements significantly. Other effects of the compound genotype included: (1) delayed shut off of early progenitor-specific genes (*c-Kit*, *Vpreb1* and *Igll1*), (2) delayed activation of stage-specific markers (*Ikzf3*, *Cd25* and *Cd2*) and (3) loss of most pre-B, immature and mature B cells, which normally express the highest levels of EBF1. The data are consistent with stochastic mechanisms requiring increased levels of EBF1 at the pro-B to pre-B cell boundary.

It is notable that $Ebf1^{+/-}Tcfe2a^{+/-}$ and $Ebf1^{+/-}Runx1^{+/-}$ compound haplo-insufficient mice have defective B cell development, while other combinations of heterozygous alleles ($Pax5^{+/-}Runx1^{+/-}$, $Ebf1^{+/-}Ikzf1^{+/-}$ and $Ebf1^{+/-}Gfi1^{+/-}$) do not exhibit compound effects in B cells (K.L. and J.H., unpublished data; H. Xu and H. Singh, privileged communication). In mammals, the expression patterns of most genes are not affected by haploinsufficiency (Qian and Zhang, 2008). Thus, it can be concluded that cooperative interactions between factors at nearby sites may provide special conditions resulting in reduced transcription in mice with compound genotypes. An exception to this model was observed at the $Cd79a$ promoter, which binds EBF1 and Runx1 with robust cooperativity. Expression of $Cd79a$ transcripts was not affected in $Ebf1^{+/-}Runx1^{+/-}$ compound haplo-insufficient mice. A high frequency of co-occupancy of promoters and enhancers by EBF1 and E2A, or by EBF1 and Runx1, has now been confirmed by genomic analysis at a wide range of loci in developing B cells.

4.3 EBF1 and B Cell Lineage Commitment

Lineage commitment is defined as the fixation of cells in a single lineage or fate with the loss of potential to generate cells of other lineages. In higher eukaryotes, mechanisms governing lineage commitment are best understood in B cells due to the discovery that Pax5 regulates this process (Nutt et al. 1999; Rolink et al. 1999; Mikkola et al. 2002). Pax5 contributes to B cell development in three major ways: Pax5 (1) regulates B cell-specific gene expression in concert with EBF1, E2A and other factors (reviewed in Ramirez et al. 2010), (2) positively regulates the expression of $Ebf1$ genes (Roessler et al. 2007) and (3) represses the expression of a large spectrum of non-B cell specific genes (Delogu et al. 2006). Together, these mechanisms help activate the B cell program while restricting the expression of other programs in committed B cells.

Recent work by Pongubala and colleagues elucidated a role for EBF1 in regulating B cell commitment (Pongubala et al. 2008). Lymphoid progenitors of $Ebf1^{-/-}$ mice are unable to differentiate into B cells and possess increased potential to generate myeloid, dendritic and NK cells in reconstituted mice. These cells also produced $CD4^+CD8^+$ double-positive and single-positive T cells in RAG-deficient mice. Increased expression of EBF1 in MPPs induced B cell development at the expense of myeloid development in vitro. Importantly, EBF1 inhibited the ability of Pax5-deficient cells to attain alternative fates. These data suggest that the expression of EBF1 is a commitment signal to the B lineage in the absence of Pax5.

Clues as to how EBF1 promotes the B cell fate were derived from the analysis of gene expression with and without EBF1. In transduced cells, EBF1 antagonized expression of both C/EBPα and PU.1, major determinants of myeloid differentiation. EBF1 also repressed expression of Id2 and Id3, which inhibit the functional activities of E2A proteins (Pongubala et al. 2008; Thal et al. 2009). $Id2$, $Id3$ and $Cebpa$ genes (encoding C/EBPα) are targets of direct regulation by EBF1.

The ability of EBF1 to repress genes characteristic of earlier progenitors and of other lineages is also dosage-dependent. Pro-B cells of $Ebf1^{+/-}$ mice expressed significantly higher levels of $Ly6a$/Sca1 than cells of wild-type mice (Lukin et al. 2011). The haplo-insufficient pro-B cells also expressed multiple NK cell-specific genes, including $Cd244$ (2B4). These effects were more apparent in these pro-B cells and occurred despite expression of Pax5. Enforced expression of EBF1 in $Ebf1^{+/-}Runx1^{+/-}$ pro-B cells increased expression of B cell markers including $Igll1$ and repressed expression of NK-lineage genes including $Cd244$, $Cd160$, $Klrb1c$ (NK1.1) and $Nfil3$ (E4BP4), a transcription factor that contributes to NK cell development (Gascoyne et al. 2009). The data suggest that normal levels of EBF1 are required for B cell identity, which EBF1 maintains by repressing the expression of non-B lineage-specific genes.

5 Identification of EBF1 Target Genes Using Genome-Wide Analysis

The full extent of EBF1's control of the B cell transcriptome was made apparent recently in two extensive studies (Lin et al. 2010; Treiber et al. 2010b). These publications made use of ChIP-based technologies to assess the breadth of EBF1 binding to regulatory modules in the chromatin of pro-B cells. Common themes arising from these experiments included the extent of EBF1 DNA binding, which was detected at >500 genes. These sites were localized within sets of genes involved in a host of biological processes including pre-BCR, BCR and PI3K/Akt/ mTOR signaling, B cell adhesion, cell cycle control and migration. The studies also correlated EBF1 DNA binding with the status of epigenetic marks on histones. Notably, EBF1 binding correlated with histone modifications necessary for appropriate transcriptional activation and repression characteristic of the B cell program. This suggests that EBF1 plays a primary role in the epigenetic regulation of B cell development.

5.1 Binding of EBF1 to a Vast Array of Sites in Pro-B Cells

Treiber et al. (2010b) used multiple methods to identify functionally important binding sites of EBF1 in vivo. First, a 'ChIP-on-chip' approach was employed to estimate the number of genes that are occupied by EBF1 in pro-B cells. Immunoprecipitated fragments were used to probe DNA tiling arrays consisting of 17,000 DNA fragments surrounding transcriptional start sites (TSS) of promoter regions. As a result, 228 potential target genes were identified. The functional importance of EBF1 binding to these genes was confirmed by comparing patterns of gene expression in $Ebf1^{-/-}$ pre-pro-B cells without and with expression of EBF1. A high percentage of potential target genes were activated by enforced

EBF1 expression in EBF1-deficient cells, including the canonical EBF1 targets *Igll1* and *Vpreb1*. A second set of experiments examined consequences of the loss of EBF1 following Cre-mediated deletion of floxed *Ebf1* alleles in gene-targeted mice. The loss of EBF1 due to Cre resulted in down-regulation of genes identified in the ChIP-on-chip experiments. Overlap between the datasets of gain-of-function and loss-of-function experiments confirmed the regulation of a subset of genes by EBF1, including *Hes1* (Notch pathway), *Sox4* and *Pou2af1*. The experiments confirmed the previous identification of *Ceacam1* and the immunomodulatory adaptor gene *Dok3* as targets of EBF1 regulation (Månsson et al. 2007; Ng et al. 2007). The list of genes repressed by EBF1 also included *Pdcd1* and *Ctla4* (members of the CD28 family), *Icosl* (the ligand of the Icos receptor) and *Hlx* (a homeobox factor that regulates Th1 differentiation). Overall, the patterns of activated and repressed genes suggest a positive role for EBF1 in regulating B cell signaling, including BCR, CD19 and the PI3K pathways, while repressing signaling pathways that are important in other cell types. It is noteworthy that many EBF1-regulated genes involved in pre-BCR and BCR signaling are co-regulated by Pax5, underscoring the importance of the EBF1-Pax5 axis in the B cell program.

To extend the analysis of EBF1 DNA binding to regions outside known promoters, Treiber et al. sequenced libraries of DNA generated following immunoprecipitation using anti-EBF1 antibodies or control input DNA. This deep sequencing approach identified $\sim$4,500 binding sites occupied by EBF1 within 100 kb of annotated genes (corresponding to 5,025 genes). The data displayed a high degree of concordance with previous ChIP-on-chip data, as over 94% of previously identified sites were also detected using deep sequencing. Interestingly, the sites were restricted to genes that are expressed specifically in B cells. Binding was not observed on genes that are thought to bind EBF1 in other cell types (i.e., forebrain neurons, adipocytes and osteoblasts). Therefore, EBF1 binding is restricted to genes of the B cell program in B cells.

5.2 Co-occupancy of Genes by EBF1 and E2A

Lin and colleagues (Lin et al. 2010) used a ChIP-seq approach not only to detect the presence of EBF1 at promoter, enhancer and intragenic regions of genes in murine pro-B cells, but also focused extensively on co-occupancy by other factors that contribute to the B cell fate (Lin et al. 2010). The compiled sequences were used to generate the consensus 5′-G/A/TC/GTCCCT/C/AA/G/TGGGA-3′, which is very similar to the optimized EBF1 site identified previously using binding site selection in vitro (5′-ATTCCCNNGGGAAT-3′)(Travis et al. 1993). In addition to detection of EBF1 DNA binding, ChIP-seq was used to localize binding sites for E2A and the Forkhead protein FOXO1 in pre-pro-B and pro-B cells. While comparative analysis detected only rare evidence of co-occupancy of sites (within 150 base pairs) by EBF1 and FOXO1. Nine percent of the 1,753 genes upregulated in pro-B cells were co-occupied by EBF1 and E2A. Interestingly, sites occupied by E2A shifted significantly between pre-pro-B cells in the absence of EBF1 and pro-B cells that express EBF1. Sites common to both E2A and EBF1 binding corresponded to genes involved in processes required for B cell development, including transcriptional regulators, cell differentiation, control of the cell cycle and signaling via the PI3K pathway, overlapping significantly with the list of (Treiber et al. 2010b). Together, these data confirm the importance of synergistic interactions between EBF1 and E2A.

5.3 Epigenetic Regulation of Genes by EBF1

DNA binding by EBF1 has been linked with changes in epigenetic marks at target genes. Lin and colleagues (Lin et al. 2010) correlated the binding of EBF1 and other factors including E2A with a series of histone modifications associated with transcriptional activation or repression. Interestingly, the presence of the E2A protein E47 correlated with histone H3K4 monomethylation (H3K4me1) at target genes, which included those co-regulated by EBF1. H3K4me1 likely constitutes a mark of 'poised' chromatin, an intermediate state that facilitates additional epigenetic modifications necessary for transcription (Robertson et al. 2008). However, at co-regulated genes, increased acquisition of the active trimethylated mark H3K4me3 and subsequent H3 acetylation was observed following DNA binding by EBF1. These data imply that active transcription requires cooperation between EBF1 and E2A.

Treiber et al. also noted differences in epigenetic marks in the absence or presence of EBF1 in B cell progenitors (Treiber et al. 2010b). Three classes of B cell-specific genes were identified that bind EBF1 at their promoters: activated, repressed and poised. At activated promoters, chromatin in the vicinity of EBF1 binding sites gained the H3K4me2 mark during the transition between pre-pro-B and pro-B cells, which may be concurrent with expression of E2A. H3K4me3 and H3 acetylation were noted in pro-B cells that expressed high levels of *Ebf1* transcripts. In contrast, promoters of repressed genes possessed variable degrees of H3K4 methylation, but were more likely to maintain negative H3K27me3 marks. Poised genes represent a group in which EBF1 modifications of chromatin, including H3K4me2, are induced at early stages and in the absence of transcription. However, EBF1-dependent transcription of these genes occurs only at later stages (mature B cells). H3K4me2 was identified previously as a mark of developmental 'poising' at hematopoietic genes (Orford et al. 2008). Transcription of these poised genes (including *Cd40*) in mature B cells is ostensibly due to the need for other factors, which require the modifications implemented by EBF1 to activate transcription at later stages of development. Interestingly, one report suggests that factors (i.e. Sox4) induce histone modifications (including H3K4me2) in the chromatin of EBF1 target genes (*Igll1-Vpreb1*) as early as blood-forming hemangioblasts (Liber et al. 2010). Together, these data have led to a revision of

B Lymphocyte Lineage Specification, Commitment and Epigenetic Control 31

the characterization of EBF1 as a pioneer factor (Hagman and Lukin, 2005) because EBF1 mediates changes in chromatin and facilitates the binding of other proteins in a 'hematopoietic context' (Treiber et al. 2010b).

5.4 Multiple Mechanisms Activate Cd79a Transcription in Early B Cells

For many years, the B cell-specific Cd79a promoter has served as a useful model for defining requirements for gene activation in early progenitors, and thus, B lineage specification. When fully activated, the TATA-less promoter (localized to ~ 200 base pairs) binds multiple lineage-restricted DNA-binding proteins including EBF1, Runx1(and its obligate partner CBFβ), E2A proteins and Pax5, which recruits Ets family proteins to bind a composite site (Fig. 3) (Hagman et al. 1991; Travis et al. 1991; Feldhaus et al. 1992; Fitzsimmons et al. 1996; Sigvardsson et al. 2002; Maier et al. 2004). Based on recent publications, it is possible to propose a new model that incorporates factor binding, histone modifications, nucleosome remodeling and DNA demethylation in the activation of Cd79a transcription.

Prior to their activation, Cd79a promoters are maintained in a highly-condensed state that is relatively inaccessible (Gao et al. 2009). Analysis of histone modifications in non-B cells and pre-pro-B cells lacking EBF1 suggests that a 'hematopoietic' state is maintained in Cd79a promoter chromatin prior to the expression of EBF1 (Treiber et al. 2010a, b). Key features of this state may include: (1) repression by Polycomb group proteins, which have been linked with repression of the Pax5 gene prior to its activation by EBF1 (Decker et al. 2009), (2) epigenetic modifications that facilitate EBF1 binding (Heinz et al. 2010; Liber et al. 2010), and/or (3) interactions with E2A, Runx1 or other transcription factors. Recent data demonstrated that E2A, which is expressed as early as HSCs, is sufficient for modifications of histones at Cd79a promoters that include H3K4me1 (Lin et al. 2010). An additional modification that has received less attention is DNA methylation, which modifies CpG dinucleotides with 5-methylcytosine (reviewed in Jaenisch and Bird, 2003; Bonasio et al. 2010). Prior to the expression of EBF1, Cd79a promoters possess only methylated CpGs, which are generally associated with inactive promoters. In turn, hypermethylated chromatin with inactive histone modifications provides a substrate for Mi-2/Nucleosome Remodeling and Deacetylase (Mi-2/NuRD) complexes. Mi-2/NuRD includes subunits that bind methylated DNA, deacetylate or demethylate histones and mobilize nucleosomes to assemble and maintain compact chromatin (reviewed in Ramírez and Hagman, 2009).

Biochemical experiments suggest that, following its expression in CLPs/pre-pro-B cells, EBF1 assembles complexes on Cd79a promoters with Runx1/CBFβ and E2A proteins. Experiments in murine EBF1-deficient fetal liver progenitors and plasmacytoma cells demonstrated that EBF1 is sufficient to initiate DNA demethylation and chromatin remodeling. These modifications increase local

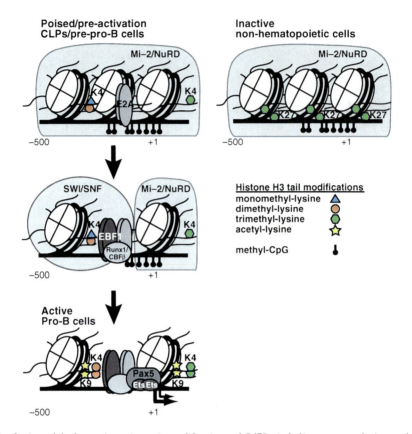

Fig. 3 *A model of stepwise epigenetic modifications of Cd79a (mb-1) promoters during early B cell development.* Transcriptional activation of the *Cd79a* gene is necessary for progression to the pre-B cell stage. Inactive *Cd79a* promoters in non-hematopoietic cells (*right*) are characterized by histone H3K27 trimethylation, a mark of heterochromatin and repression. Chromatin at these promoters is likely maintained in an inactive state by repressive Mi-2/NuRD chromatin remodeling complexes (CRCs). During pre-activation of *Cd79a* genes in common lymphoid progenitors (CLPs) and pre-pro-B cells (*top left*), H3K4 mono- and di-methylation is correlated with binding of the transcription factor E2A to *Cd79a* promoters. Poised promoters generally display low levels of mono-, di- and tri-methylated H3K4 and do not display significant levels of H3K27 trimethylation. Binding of EBF1 initiates demethylation of promoter DNA and recruits activating SWI/SNF CRCs, which result in nucleosome displacement (*middle left*). Activation of *Cd79a* expression in pro-B cells involves the subsequent recruitment of other transcription factors including Pax5 and Ets proteins and correlates with **a** displacement of repressive Mi-2/NuRD complexes, **b** complete demethylation of promoter DNA, **c** increased H3K4 di- and trimethylation with loss of monomethylation and **d** high levels of H3K9 acetylation (*bottom left*). This model is based, in part, on data from Gao et al. (2009), Lin et al. (2010), Treiber et al. (2010b) and unpublished data (J.R. and J.H.)

chromatin accessibility, but they do not lead to transcription in the absence of other transcription factors (Gao et al. 2009). In both of these cellular models, expression of Pax5 was important for the propagation of DNA demethylation and subsequent transcription. These processes are dependent on SWI/SNF chromatin remodeling complexes, which are recruited to the promoter following DNA binding by EBF1. The activity of SWI/SNF is opposed functionally by Mi-2/NuRD. shRNA-mediated depletion of Mi-2β (*Chd4*), a core catalytic subunit of Mi-2/NuRD, greatly enhances activation of *Cd79a* promoters by EBF1 and Pax5 in plasmacytoma cells. Depleting Mi-2β facilitates the propagation of DNA demethylation and allows for the assembly of Pax5:Ets ternary complexes on unmethylated CpGs (Maier et al. 2003; Maier et al. 2004). The mechanism by which EBF1, Pax5 and SWI/SNF exclude Mi-2/NuRD from the *Cd79a* promoter is unknown. A dynamic competition likely exists between EBF1 and other transcription factors that promote DNA demethylation versus the recruitment of Mi-2/NuRD to methylated CpGs.

Following the binding of *Cd79a* promoters by EBF1, Pax5 and other proteins, additional changes in histone modifications take place. These activities may be mediated by histone acetyltransferase (HAT) domains of the p300 co-activator, which are recruited by EBF1 (Zhao et al. 2003; He et al. 2011). Interestingly, p300 can acetylate Pax5 directly, which increases its activity in some contexts (He et al. 2011). Efficient transcription is associated with a reduction in nucleosome density across the promoter region (Gao et al. 2009; J.R. and J.H., data not shown) and an increase in histone H3K4 trimethylation and H3 acetylation at flanking nucleosomes (Lin et al. 2010; Treiber et al. 2010b). Although much has been revealed recently, additional experiments are needed to determine the identities and define the temporal progression of factors that mediate the changes in chromatin structure necessary for *Cd79a* transcription.

6 Conclusions

Initial investigations concerning EBF1 confirmed its regulation of key genes necessary for early B cell development. These observations led to the hypothesis that EBF1 participates in the specification and commitment of B cell progenitors. It is now established that EBF1 instructs this lineage choice. In collaboration with a hierarchy of partner proteins, including E2A, Runx1 and Pax5, EBF1 activates the B cell transcriptome and represses programs of alternate hematopoietic lineages.

The elucidation of much of the structure of EBF1 provides important information concerning its DNA binding, and provides clues to how it may interact with other proteins and activate transcription. The structures provide a basis for addressing aspects of its regulation, including the potential for modulating its activity and integration with signaling pathways. Additional structural and functional studies are needed to elucidate requirements for the three α-helices in the HLHLH domain of EBF1, while only two α-helices suffice for functions of b-HLH proteins. The role of the TIG/IPT domain also remains an enigma. Does it enhance dimerization or does it have other functions?

Collaborative networks of transcription factors direct lineage specification in all known developmental programs. Data confirms the essential role of EBF1 in the B cell network. As early as the CLP stage, EBF1 directs the expression of genes of the B cell program. Failure to express sufficient levels of EBF1 results in reduced ability to maintain commitment and limits production of B cells. Although it has been determined that EBF1 regulates hundreds of genes in early B cells, many questions remain. At early stages of development, how are different outcomes of transcriptional activation or repression achieved by EBF1? EBF1 activates genes largely in concert with E2A. Do interactions with other factors convert EBF1 into a repressor? Does EBF1 regulate V(D)J recombination directly?

Recent studies have established multiple roles for EBF1 in modulating chromatin structure. However, the underlying mechanisms are yet to be revealed. Does EBF1 recruit histone modifying enzymes and/or chromatin remodeling complexes directly? How does EBF1 facilitate the demethylation of DNA? Are Mi-2/NuRD complexes excluded from promoters by EBF1?

Regulation of early B cell development by EBF1 has provided a paradigm for the control of transcription and cellular differentiation. EBF1 has also proven to be an important regulator of signaling pathways in B cells. Future investigations should address the functions of EBF1 in B cell responses to antigens.

Acknowledgments J.R. was supported by NIH Post-doctoral Training Grant T32 AI07405. J.H. is generously supported by NIH grants R01 AI54661, R01 AI81878 and P01 AI22295.

References

Adolfsson J, Mansson R, Buza-Vidas N, Hultquist A, Liuba K, Jensen CT, Bryder D, Yang L, Borge OJ, Thoren LA, Anderson K, Sitnicka E, Sasaki Y, Sigvardsson M, Jacobsen SE (2005) Identification of Flt3 + lympho-myeloid stem cells lacking erythro-megakaryocytic potential a revised road map for adult blood lineage commitment. Cell 121:295–306
Aravind L, Koonin E (1999) Gleaning non-trivial structural, functional and evolutionary information about proteins by iterative database searches. J Mol Biol 287:1023–1040
Bain G, Maandag EC, Izon DJ, Amsen D, Kruisbeek AM, Weintraub BC, Krop I, Schlissel MS, Feeney AJ, van Roon M, van der Valk M, te Reile HPJ, Berns A, Murre C (1994) E2A proteins are required for proper B cell development and initiation of immunoglobulin gene rearrangements. Cell 79:885–892
Bhalla S, Spaulding C, Brumbaugh RL, Zagort DE, Massari ME, Murre C, Kee BL (2008) differential roles for the E2A activation domains in B lymphocytes and macrophages. J Immunol 180:1694–1703
Bonasio R, Tu S, Reinberg D (2010) Molecular signals of epigenetic states. Science 330:612–616
Campbell KS, Hager EJ, Friedrich RJ, Cambier JC (1991) IgM antigen receptor complex contains phosphoprotein products of B29 and mb-1 genes. Proc Natl Acad Sci USA 88:3982–3986
Decker T, Pasca di Magliano M, McManus S, Sun Q, Bonifer C, Tagoh H, Busslinger M (2009) Stepwise activation of enhancer and promoter regions of the B cell commitment gene Pax5 in early lymphopoiesis. Immunity 30:508–520
DeKoter RP, Singh H (2000) Regulation of B lymphocyte and macrophage development by graded expression of PU.1. Science 288:1439–1441
Delogu A, Schebesta A, Sun Q, Aschenbrenner K, Perlot T, Busslinger M (2006) Gene repression by Pax5 in B cells is essential for blood cell homeostasis and is reversed in plasma cells. Immunity 24:269–281

B Lymphocyte Lineage Specification, Commitment and Epigenetic Control

Dengler HS, Baracho GV, Omori SA, Bruckner S, Arden KC, Castrillon DH, DePinho RA, Rickert RC (2008) Distinct functions for the transcription factor Foxo1 at various stages of B cell differentiation. Nat Immunol 9:1388–1398

Dias S, Silva H Jr, Cumano A, Viera P (2005) Interleukin-7 is necessary to maintain the B cell potential in common lymphoid progenitors. J Exp Med 201:971–979

Dias S, Mansson R, Gurbuxani S, Sigvardsson M, Kee BL (2008) E2A proteins promote development of lymphoid-primed multipotent progenitors. Immunity 29:217–227

Ellenberger T, Fass D, Arnaud M, Harrison SC (1994) Crystal structure of transcription factor E47: E box recognition by a basic region helix-loop-helix dimer. Genes Dev 8:970–980

Feldhaus A, Mbangkollo D, Arvin K, Klug C, Singh H (1992) BlyF, a novel cell-type- and stage-specific regulator of the B-lymphocyte gene *mb-1*. Mol Cell Biol 12:1126–1133

Fields S, Ternyak K, Gao H, Ostraat R, Akerlund J, Hagman J (2008) The 'zinc knuckle' motif of Early B cell Factor is required for transcriptional activation of B cell-specific genes. Mol Immunol 45:3786–3796

Finkler A, Ashery-Padan R, Fromm H (2007) CAMTAs: calmodulin-binding transcription activators from plants to human. FEBS Lett 581:3893–3898

Fitzsimmons D, Hodsdon W, Wheat W, Maira S-M, Wasylyk B, Hagman J (1996) Pax-5 (BSAP) recruits Ets proto-oncogene family proteins to form functional ternary complexes on a B-cell-specific promoter. Genes Dev 10:2198–2211

Gao H, Lukin K, Ramírez J, Fields S, Lopez D, Hagman J (2009) Opposing effects of SWI/SNF and Mi-2/NuRD chromatin remodeling complexes on epigenetic reprogramming by EBF and Pax5. Proc Natl Acad Sci USA 106:11258–11263

Gascoyne DM, Long E, Veiga-Fernandes H, de Boer J, Williams O, Seddon B, Coles M, Kioussis D, Brady HJ (2009) The basic leucine zipper transcription factor E4BP4 is essential for natural killer cell development. Nat Immunol 10:1118–1124

Goebel P, Janney N, Valenzuela JR, Romanow WJ, Murre C, Feeney AJ (2001) Localized gene-specific induction of accessibility to V(D)J recombination induced by E2A and early B cell factor in nonlymphoid cells. J Exp Med 194:645–656

Gold MR, Matsuuchi L, Kelly RB, DeFranco AL (1991) Tyrosine phosphorylation of components of the B-cell antigen receptors following receptor crosslinking. Proc Natl Acad Sci USA 88:3436–3440

Hagman J, Lukin K (2005) Early B cell factor 'pioneers' the way to B cell development. Trends Immunol 26:455–461

Hagman J, Travis A, Grosschedl R (1991) A novel lineage-specific nuclear factor regulates mb-1 gene transcription at the early stages of B cell differentiation. EMBO J 10:3409–3417

Hagman J, Belanger C, Travis A, Turck CW, Grosschedl R (1993) Cloning and functional characterization of early B-cell factor, a regulator of lymphocyte-specific gene expression. Genes Dev 7:760–773

Hagman J, Gutch MJ, Lin H, Grosschedl R (1995) EBF contains a novel zinc coordination motif and multiple dimerization and transcriptional activation domains. EMBO J 14:2907–2916

Hardy RR, Kincade PW, Dorshkind K (2007) The protean nature of cells in the B lymphocyte lineage. Immunity 26:703–714

He T, Hong SY, Huang L, Xue W, Yu Z, Kwon H, Kirk M, Ding SJ, Su K, Zhang Z (2011) Histone acetyltransferase p300 acetylates Pax5 and strongly enhances Pax5-mediated transcriptional activity. J Biol Chem 286:14137–14145

Heinz S, Benner C, Spann N, Bertolino E, Lin YC, Laslo P, Cheng JX, Murre C, Singh H, Glass CK (2010) Simple combinations of lineage-determining transcription factors prime cis-regulatory elements required for macrophage and B cell identities. Mol Cell 38:576–589

Herzog S, Hug E, Meixlsperger S, Paik JH, DePinho RA, Reth M, Jumaa H (2008) SLP-65 regulates immunoglobulin light chain gene recombination through the PI(3)K-PKB-Foxo pathway. Nat Immunol 9:623–631

Hess J, Nielsen PJ, Fischer KD, Bujard H, Wirth T (2001) The B lymphocyte-specific coactivator BOB.1/OBF.1 is required at multiple stages of B-cell development. Mol Cell Biol 21:1531–1539

Hombach J, Lottspeich F, Reth M (1990) Identification of the genes encoding the IgM-alpha and Ig-beta components of the IgM antigen receptor complex by amino-terminal sequencing. Eur J Immunol 20:2795–2799

Ichii M, Shimazu T, Welner RS, Garrett KP, Zhang Q, Esplin BL, Kincade PW (2010) Functional diversity of stem and progenitor cells with B-lymphopoietic potential. Immunol Rev 237:10–21

Jaenisch R, Bird A (2003) Epigenetic regulation of gene expression: how the genome integrates intrinsic and environmental signals. Nat Genet 33(Suppl):245–254

Kee BL, Murre C (1998) Induction of early B cell factor (EBF) and multiple B lineage genes by the basic helix-loop-helix transcription factor E12. J Exp Med 188:699–713

Kikuchi K, Lai AY, Hsu C-L, Kondo M (2005) IL-7 receptor signaling is necessary for stage transition in adult B cell development through up-regulation of EBF. J Exp Med 201:1197–1203

Kim U, Qin X-F, Gong S, Stevens S, Luo Y, Nussenzweig M, Roeder RG (1996) The B-cell-specific transcription coactivator OCA-B/OBF-1/Bob-1 is essential for normal production of immunoglobulin isotypes. Nature 383:542–547

Klug A (2010) The discovery of zinc fingers and their applications in gene regulation and genome manipulation. Annu Rev Biochem 79:213–231

Kondo M (1997) Identification of clonogenic lymphoid progenitors in mouse bone marrow. Cell 91:661–672

Kondo M (2010) Lymphoid and myeloid lineage commitment in multipotent hematopoietic progenitors. Immunol Rev 238:37–46

Liber D, Domaschenz R, Holmqvist PH, Mazzarella L, Georgiou A, Leleu M, Fisher AG, Labosky PA, Dillon N (2010) Epigenetic priming of a pre-B cell-specific enhancer through binding of Sox2 and Foxd3 at the ESC stage. Cell Stem Cell 7:114–126

Lin H, Grosschedl R (1995) Failure of B-cell differentiation in mice lacking the transcription factor EBF. Nature 376:263–267

Lin YC, Jhunjhunwala S, Benner C, Heinz S, Welinder E, Månsson R, Sigvardsson M, Hagman J, Espinoza CA, Dutkowski J, Ideker T, Glass CK, Murre C (2010) A global network of transcription factors, involving E2A, EBF1 and Foxo1, that orchestrates B cell fate. Nat Immunol 11:635–643

Lukin K, Fields S, Lopez D, Cherrier M, Ternyak K, Ramirez J, Feeney AJ, Hagman J (2010) Compound haploinsufficiencies of Ebf1 and Runx1 genes impede B cell lineage progression. Proc Natl Acad Sci USA 107:7869–7874

Lukin K, Fields S, Guerrettaz L, Straign D, Rodriguez V, Zandi S, Mansson R, Cambier JC, Sigvardsson M, Hagman J (2011) A dose-dependent role for EBF1 in repressing non-B-cell-specific genes. Eur J Immunol 41:1787–1793

Maier H, Colbert J, Fitzsimmons D, Clark DR, Hagman J (2003) Activation of the early B cell-specific mb-1 (Ig-a) gene by Pax-5 is dependent on an unmethylated Ets binding site. Mol Cell Biol 23:1946–1960

Maier H, Ostraat R, Gao H, Fields S, Shinton SA, Medina KL, Ikawa T, Murre C, Singh H, Hardy RR, Hagman J (2004) Early B cell factor cooperates with Runx1 and mediates epigenetic changes associated with mb-1 transcription. Nat Immun 5:1069–1077

Malin S, McManus S, Cobaleda C, Novatchkova M, Delogu A, Bouillet P, Strasser A, Busslinger M (2010) Role of STAT5 in controlling cell survival and immunoglobulin gene recombination during pro-B cell development. Nat Immunol 11:171–179

Mansson R, Zandi S, Welinder E, Tsapogas P, Sakaguchi N, Bryder D, Sigvardsson M (2010) Single-cell analysis of the common lymphoid progenitor compartment reveals functional and molecular heterogeneity. Blood 115:2601–2609

Månsson R, Lagergren A, Hansson F, Smith E, Sigvardsson M (2007) The CD53 and CEACAM-1 genes are genetic targets for early B cell factor. Eur J Immunol 37:1365–1376

Medina KL, Pongubala JMR, Reddy KL, Lancki DW, DeKoter RP, Kieslinger M, Grosschedl R, Singh H (2004) Defining a regulatory network for specification of the B cell fate. Dev Cell 7:607–617

Mikkola I, Heavey B, Horcher M, Busslinger M (2002) Reversion of B cell commitment upon loss of Pax5 expression. Science 297:110–113

Ng CH, Xu S, Lam KP (2007) Dok-3 plays a nonredundant role in negative regulation of B-cell activation. Blood 110:259–266

Ng SY, Yoshida T, Zhang J, Georgopoulos K (2009) Genome-wide lineage-specific transcriptional networks underscore Ikaros-dependent lymphoid priming in hematopoietic stem cells. Immunity 30:493–507

Nutt SL, Heavey B, Rolink AG, Busslinger M (1999) Commitment to the B-lymphoid lineage depends on the transcription factor Pax5. Nature 401:556–562

O'Riordan M, Grosschedl R (1999) Coordinate regulation of B cell differentiation by the transcription factors EBF and E2A. Immunity 11:21–31

Orford K, Kharchenko P, Lai W, Dao MC, Worhunsky DJ, Ferro A, Janzen V, Park PJ, Scadden DT (2008) Differential H3K4 methylation identifies developmentally poised hematopoietic genes. Dev Cell 14:798–809

Pongubala JM, Northrup DL, Lancki DW, Medina KL, Treiber T, Bertolino E, Thomas M, Grosschedl R, Allman D, Singh H (2008) Transcription factor EBF restricts alternative lineage options and promotes B cell fate commitment independently of Pax5. Nat Immunol 9:203–215

Purohit SJ, Stephan RP, Kim HG, Herrin BR, Gartland L, Klug CA (2003) Determination of lymphoid cell fate is dependent on the expression status of the IL-7 receptor. EMBO J 22: 5511–5521

Qian W, Zhang J (2008) Gene dosage and gene duplicability. Genetics 179:2319–2324

Ramirez J, Hagman J (2009) The Mi-2/NuRD complex: a critical epigenetic regulator of hematopoietic development, differentiation and cancer. Epigenetics 4:532–536

Ramirez J, Lukin K, Hagman J (2010) From hematopoietic progenitors to B cells: mechanisms of lineage restriction and commitment. Curr Opin Immunol 22:177–184

Reynaud D, Demarco IA, Reddy KL, Schjerven H, Bertolino E, Chen Z, Smale ST, Winandy S, Singh H (2008) Regulation of B cell fate commitment and immunoglobulin heavy-chain gene rearrangements by Ikaros. Nat Immunol 9:927–936

Robertson AG, Bilenky M, Tam A, Zhao Y, Zeng T, Thiessen N, Cezard T, Fejes AP, Wederell ED, Cullum R, Euskirchen G, Krzywinski M, Birol I, Snyder M, Hoodless PA, Hirst M, Marra MA, Jones SJ (2008) Genome-wide relationship between histone H3 lysine 4 mono- and tri-methylation and transcription factor binding. Genome Res 18:1906–1917

Roessler S, Györy I, Imhof S, Spivakov M, Williams RR, Busslinger M, Fisher AG, Grosschedl R (2007) Distinct promoters mediate the regulation of Ebf1 gene expression by interleukin-7 and Pax5. Mol Cell Biol 27:579–594

Rolink AG, Nutt SL, Melchers F, Busslinger M (1999) Long-term in vivo reconstitution of T-cell development by Pax5-deficient B-cell progenitors. Nature 401:603–606

Romanow WJ, Langerak AW, Goebel P, Wolvers-Tettero ILM, van Dongen JJM, Feeney AJ, Murre C (2000) E2A and EBF act in synergy with the V(D)J recombinase to generate a diverse immunoglobulin repertoire in nonlymphoid cells. Mol Cell 5:343–353

Rothenberg EV (2007) Cell lineage regulators in B and T cell development. Nat Immunol 8:441–444

Schubart DB, Rolink A, Kosco-Vilbois MH, Botteri F, Matthias P (1996) B-cell-specific coactivator OBF-1/OCA-B/Bob1 required for immune response and germinal centre formation. Nature 383:538–542

Semerad CL, Mercer EM, Inlay MA, Weissman IL, Murre C (2009) E2A proteins maintain the hematopoietic stem cell pool and promote the maturation of myelolymphoid and myeloerythroid progenitors. Proc Natl Acad Sci USA 106:1930–1935

Sigvardsson M, O'Riordan M, Grosschedl R (1997) EBF and E47 collaborate to induce expression of the endogenous immunoglobulin surrogate light chain genes. Immunity 7:25–36

Sigvardsson M, Clark DR, Fitzsimmons D, Doyle M, Åkerblad P, Breslin T, Bilke S, Liu Y-F, Yeamans C, Zhang G, Hagman J (2002) Early B cell Factor, E2A, and Pax-5 cooperate to activate the early B cell-specific *mb-1* promoter. Mol Cell Biol 22:8539–8551

Siponen MI, Wisniewska M, Lehtio L, Johansson I, Svensson L, Raszewski G, Nilsson L, Sigvardsson M, Berglund H (2010) Structural determination of functional domains in early B-cell factor (EBF) family of transcription factors reveals similarities to Rel DNA-binding proteins and a novel dimerization motif. J Biol Chem 285:25875–25879

Smith EMK, Gisler R, Sigvardsson M (2002) Cloning and characterization of a promoter flanking the Early B cell Factor (EBF) gene indicates roles for E-proteins and autoregulation in the control of EBF expression. J Immunol 169:261–270

Spooner CJ, Cheng JX, Pujadas E, Laslo P, Singh H (2009) A recurrent network involving the transcription factors PU.1 and Gfi1 orchestrates innate and adaptive immune cell fates. Immunity 31:576–586

Srinivasan L, Sasaki Y, Calado DP, Zhang B, Paik JH, DePinho RA, Kutok JL, Kearney JF, Otipoby KL, Rajewsky K (2009) PI3 kinase signals BCR-dependent mature B cell survival. Cell 139:573–586

Thal M, Carvalho TL, He T, Kim HG, Gao H, Hagman J, Klug CA (2009) Ebf1-mediated down-regulation of Id2 and Id3 is essential for specification of the B cell lineage. Proc Natl Acad Sci USA 106:552–557

Travis A, Hagman J, Grosschedl R (1991) Heterogeneously initiated transcription from the pre-B- and B-cell-specific *mb*-1 promoter: analysis of the requirement for upstream factor-binding sites and initiation site sequences. Mol Cell Biol 11:5756–5766

Travis A, Hagman J, Hwang L, Grosschedl R (1993) Purification of early-B-cell factor and characterization of its DNA- binding specificity. Mol Cell Biol 13:3392–3400

Treiber N, Treiber T, Zocher G, Grosschedl R (2010a) Structure of an Ebf1:DNA complex reveals unusual DNA recognition and structural homology with Rel proteins. Genes Dev 24:2270–2275

Treiber T, Mandel EM, Pott S, Györy I, Firner S, Liu ET, Grosschedl R (2010b) Early B cell factor 1 regulates B cell gene networks by activation, repression, and transcription-independent poising of chromatin. Immunity 32:714–725

Wang MM, Reed RR (1993) Molecular cloning of the olfactory neuronal transcription factor OLF-1 by genetic selection in yeast. Nature 364:121–126

Zandi S, Mansson R, Tsapogas P, Zetterblad J, Bryder D, Sigvardsson M (2008) EBF1 is essential for B-lineage priming and establishment of a transcription factor network in common lymphoid progenitors. J Immunol 181:3364–3372

Zhang Z, Cotta CV, Stephan RP, deGuzman CG, Klug CA (2003) Enforced expression of EBF in hematopoietic stem cells restricts lymphopoiesis to the B cell lineage. EMBO J 22:4759–4769

Zhao F, McCarrick-Walmsley R, Akerblad P, Sigvardsson M, Kadesch T (2003) Inhibition of p300/CBP by early B-cell factor. Mol Cell Biol 23:3837–3846

Zhuang Y, Soriano P, Weintraub H (1994) The helix-loop-helix gene E2A is required for B cell formation. Cell 79:875–884

Epigenetic Features that Regulate IgH Locus Recombination and Expression

Ramesh Subrahmanyam and Ranjan Sen

Abstract Precisely regulated rearrangements that yield imprecise recombination junctions are hallmarks of antigen receptor gene assembly. At the immunoglobulin heavy chain (IgH) gene locus this is initiated by rearrangement of a D_H gene segment to a J_H gene segment to generate DJ_H junctions, followed by rearrangement of a V_H gene segment to the DJ_H junction to generate fully recombined VDJ alleles. In this review we discuss the regulatory features of each step of IgH gene assembly and the role of epigenetic mechanisms in achieving regulatory precision.

Contents

1	Overview of the Immunoglobulin Heavy Chain Genomic Organization	40
2	The First Step of IgH Gene Assembly: D_H to J_H Recombination	40
	2.1 Features of D_H to J_H Recombination	40
	2.2 Regulation of D_H Recombination	42
	2.3 Epigenetic Features of the D_H–$C\mu$ Domain	43
	2.4 Chromosome Conformation of the D_H–$C\mu$ Domain	44
	2.5 Functional Implications of the Structure of the 3' IgH Domain	46
3	The Second Step of IgH Gene Assembly: V_H to DJ_H Recombination	48
	3.1 Chromatin Structural Changes that Accompany D_H to J_H Recombination	49
	3.2 How does DJ_H Recombination Activate V_H Recombination?	51
	3.3 Regulation of V_H Gene Recombination	51
	3.4 Differential Regulation of Proximal and Distal V_H Recombination	52

R. Subrahmanyam · R. Sen (✉)
Gene Regulation Section, Laboratory of Molecular Biology and Immunology,
National Institute on Aging, National Institutes of Health,
251 Bayview Boulevard, Baltimore, MD 21224, USA
e-mail: rs465z@nih.gov

Current Topics in Microbiology and Immunology (2012) 356: 39–63
DOI: 10.1007/82_2011_153
© Springer-Verlag Berlin Heidelberg 2011
Published Online: 21 July 2011

3.5	Epigenetic Features that Could Mediate Differential V_H Gene Recombination	55
3.6	Coordinating Locus Compaction and Histone Modification of V_H Genes	57
4	Conclusions	58
References		59

1 Overview of the Immunoglobulin Heavy Chain Genomic Organization

Immunoglobulin heavy chain (IgH) genes are assembled by somatic recombination of variable (V_H), diversity (D_H) and joining (J_H) gene segments that are spread over 2.5 Mb of the genome. The murine IgH locus contains approximately 150 V_H gene segments which, based on sequence homologies, are sub-divided into several gene families. The most prominent of these are indicated in Fig. 1. V_HJ558 and V_H3609 gene families overlap at the $5'$ end of the locus and contain the largest numbers of V_H gene segments (Johnston et al. 2006). At the $3'$ end of the V_H locus is the V_H7183 family, which includes the $3'$-most V_H gene segment V_H81X that recombines at high frequency during fetal B lymphopoiesis (Yancopoulos et al. 1984; Perlmutter et al. 1985; Lawler et al. 1987; Jeong and Teale 1988; Malynn et al. 1990; ten Boekel et al. 1997). Approximately 100 kb separate the V_H locus from a cluster of D_H gene segments. The number of D_H gene segments varies between 9 and 12 according to the mouse strain. The $5'$-most D_H gene segment, DFL16.1, and the several *DSP* gene segments after it form a part of a repeated DNA unit (Bolland et al. 2007; Chakraborty et al. 2007); repeats that lie in the middle of the cluster are more similar to each other than the repeats that lie at either flank. Beyond the repeated structure, the mouse genome contains one *DST4* gene segment and then, after a gap of 18 kb, the $3'$-most D_H gene segment, DQ52. DQ52 is located less than a kilobase from the first J_H gene segment. This organization, whereby one D_H gene segment lies very close to J_H gene segments and others lie further away in repeat structure, is found in most mammalian species for which IgH locus sequence is available (Subrahmanyam and Sen 2010). In mice, there is another DST4-like gene segment located approximately 60 kb $5'$ of DFL16.1; however, it appears to be used infrequently (Ye 2004).

2 The First Step of IgH Gene Assembly: D_H to J_H Recombination

2.1 Features of D_H to J_H Recombination

Ig gene assembly is initiated by the recombination of a D_H gene segment to one of the four J_H gene segments. D_H gene segments are flanked on both sides by recombination signal sequences (RSSs) that contain 12 bp spacers, whereas J_H

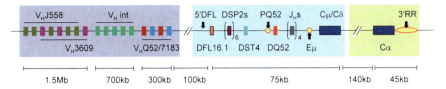

Fig. 1 Organization of the mouse immunoglobulin heavy chain gene locus. Schematic representation of the approximately 3 Mb long germline heavy chain gene locus in mouse (not to scale). The variable gene segments (V_H) occupy approximately 2.5 Mb at the 5' end of the locus (*purple (left) shaded area*) and are divided into distal (V_HJ558 and V_H3609), intermediate and proximal (V_HQ52 and V_H7183) gene families. The distal genes cover approximately 1.5 Mb of the locus with overlapping V_HJ558 and V_H3609 gene segments. The 12 intermediate V_H gene families (V_H int) span about 700 kb; each family consists 1–6 gene segments. The proximal gene families cover about 300 kb, and contain overlapping segments from V_HQ52 and V_H7183 families. The D_H–$C\mu$ part of the locus (*blue (middle) shaded area*) occupies approximately 75 kb, with the D_H region consisting of the 5' most gene segment DFL16.1, intermediate DSP and DST4 gene segments and the 3' most gene segment, DQ52. There are four J_H gene segments followed by the intronic enhancer $E\mu$ and the exons coding for constant heavy chains $C\mu$ and $C\delta$. The intronic enhancer $E\mu$ and the promoter upstream of DQ52 (PQ52) are shown as *yellow circles*. The rest of the constant region spans about 200 kb (*yellow (right) shaded area*) and consists of coding exons for six other constant heavy chain isotypes and the 3' regulatory region (3'RR, *yellow oval*) which includes the 3' Eα enhancer and seven DNase I hypersensitive sites. Looping sites at 5'DFL, PQ52, $E\mu$ and 3'RR are shown by *black arrows*

RSSs contain a 23 bp spacer. Thus, in principle, D_H to J_H recombination can proceed with inversion or deletion of the intervening DNA. Yet, in the majority of fully recombined VDJ alleles, the DJ_H junction is generated by deletional recombination (Gauss and Lieber 1992). Wu et al. estimated the frequency of inversional D_H to J_H recombination to be less than 0.1% in the adult bone marrow (Sollbach and Wu 1995). Although these observations suggest that deletion is the preferred mode of D_H recombination, the idea is based on the analysis of functional VDJ alleles and not from direct observation of primary D_H rearrangements. Therefore, it is possible that the results are skewed by other selection events. Lieber et al. analyzed the compatibility of 5' and 3' D_H-associated RSSs for recombination to the J_H RSS in the absence of additional confounding regulatory variables (Pan et al. 1997). Using recombination substrates that allowed them to quantify the level of recombination by either deletion or inversion in transient transfection assays, they found that the recombination efficiency of the 5'-RSS of D_H gene segments (that would be used in inversional recombination) is approximately tenfold lower than the 3'-RSS (that would be used in deletional recombination). While these observations provide a partial explanation for deletional preference during D_H to J_H recombination, it remains unclear why deletion is the favored route for this step of recombination. We have previously proposed that D_H recombination by deletion may activate secondary V_H recombination by reducing the distance between the V_H locus and the D_H–$C\mu$ part of the locus on DJ_H recombined alleles (Chowdhury and Sen 2001, 2004). However, the importance of deletional D_H recombination to V_H activation has not been experimentally tested yet.

Another feature of D_H to J_H recombination is that the two flanking gene segments, DFL16.1 and DQ52, are used more frequently in VDJ junctions than the numerically larger number of DSP gene segments that lie in between (Feeney 1990; Tsukada et al. 1990; Bangs et al. 1991; Chang et al. 1992; Atkinson et al. 1994; Nitschke et al. 2001). Whilst the simplest interpretation of these observations is that flanking D_H gene segments recombine more efficiently, it cannot be ruled out that the fully rearranged alleles analyzed in these studies were selected during B cell development. Furthermore, the snapshot view of functional alleles also does not account for secondary D_H–J_H rearrangements that involve the joining of D_H gene segments upstream of the DJ_H junction to downstream J_Hs. Such rearrangements would also skew the distribution of DJ_H junctions towards greater utilization of DFL16.1 without invoking greater recombination efficiency for this gene segment. Tsukada et al. (1990) carried out a quantitative analysis of the efficiency of D_H recombination in a temperature sensitive Abelson virus transformed B cell line. By restricting their analysis to primary D_H recombination, they concluded that the two flanking gene segments recombined preferentially. Our studies of D_H utilization after transient RAG2 expression in a RAG2-deficient cell line also reached the same conclusion (R. Subrahmanyam and R. Sen, unpublished observations). Thus, it is likely that the two flanking D_H gene segments are inherently more recombinogenic. Preferential rearrangement of DFL16.1 and DQ52 also ensures maximal diversity of D_H usage, which would otherwise be dominated by the more numerous, but less diverse, DSP gene segments.

2.2 Regulation of D_H Recombination

The basis for selecting gene segments for recombination is generally viewed in the context of the accessibility hypothesis of Alt and colleagues (Yancopoulos and Alt 1985; Cobb et al. 2006; Abarrategui and Krangel 2009). Its central tenet is that the recombinase gains access to different loci in cells depending on their lineage and developmental stage. For example, IgH genes are accessible to the recombinase only in B cells, but not T cells; similarly, the IgH locus is accessible to the recombinase earlier in B cell development compared to Ig light chain loci. The earliest studies of the regulation of recombination pointed to a role for chromatin structure based on the observation that accessible gene segments were sensitive to digestion by DNase I (Yancopoulos and Alt 1985; Blackwell et al. 1986; Ferrier et al. 1989). However, the nature of chromatin structural changes that confer recombinase accessibility have only recently begun to be clarified. Studies in transgenic and knockout mice show that accessibility is controlled by transcriptional regulatory sequences such as promoters and enhancers (Thomas et al. 2009). The evolving idea is that these sequences alter the epigenetic state of antigen receptor loci making them permissive for recombination. These epigenetic changes include histone modifications, chromatin remodeling, DNA methylation, sterile transcription and chromosome conformation.

2.3 Epigenetic Features of the D_H–$C\mu$ Domain

There are several striking features of the epigenetic landscape of the D_H–$C\mu$ domain of the IgH locus that may regulate D_H recombination. First, this region contains five tissue-specific DNase I hypersensitive sites (DHSs). Two of these correspond to the transcription enhancer $E\mu$ located in the J_H–$C\mu$ intron and a promoter located $5'$ of DQ52 (PQ52) where sterile μ_o transcripts initiate (Chowdhury and Sen 2001). The PQ52 DHS is abolished in the absence of $E\mu$, indicating that it is $E\mu$-dependent (Chakraborty et al. 2009). This dependence likely reflects $E\mu$-dependent transcriptional activation of PQ52. Recently, three new DHSs have been identified within 10 kb $5'$ of DFL16.1 (Featherstone et al. 2010). These DHSs contain binding sites for the transcription factor CTCF and two of the sites have been shown to have insulator activity. Based on these observations, these DHSs have been proposed to mark the $5'$ boundary of the D_H–$C\mu$ domain. Finally, the region between $E\mu$ and PQ52, that contains DQ52 and the J_HS, is much more sensitive to DNase I without actually containing DHSs (Maes et al. 2006; Chakraborty et al. 2009). The basis for accentuated DNase I sensitivity of this region is not understood.

Second, the histone modification pattern within the D_H–$C\mu$ region is markedly heterogeneous. Amongst D_H gene segments, activation-associated histone modifications, such as H3 and H4 acetylation, and H3K4-dimethylation, are restricted to DFL16.1 and DQ52. The majority of intervening DSP gene segments are, instead, associated with the repressive modification of H3K9-dimethylation. Thus, most of the 50 kb D_H cluster has features of heterochromatin. We have previously proposed that heterochromatinization of this region may be initiated at the DSP repeats by a mechanism analogous to centromeric heterochromatin formation in *S. Pombe* (Chakraborty et al. 2007; Subrahmanyam and Sen 2010). Consistent with the presence of repressive histone modifications in this region, DSP gene segments are also relatively insensitive to DNase I digestion compared to DFL16.1 and DQ52 (Chakraborty et al. 2009). Because these distinct differences in chromatin state between flanking (DFL16.1 and DQ52) and intervening (DSP and DST4) D_H gene segments may account for the differences in their recombination efficiency, it is important to understand how the state is generated and to identify the *cis*-regulatory sequences and transcription factors involved. Interestingly, the highly DNase I sensitive region encompassing the J_H gene segments is also marked by the highest levels of acetylated and H3K4-trimethylated histones compared to the rest of the D_H–$C\mu$ domain. The unique characteristics of this small domain are likely to be important for IgH gene rearrangements and expression.

Third, tissue-specific DNA de-methylation is found only at PQ52 and $E\mu$, thereby correlating with the two best characterized DHSs (R. Selimyan and R. Sen, unpublished observations). The region surrounding the DHSs upstream of DFL16.1 has not been systematically analyzed for DNA methylation, thus leaving open the question of whether all DHSs in the D_H–$C\mu$ domain are targeted for hypomethylation. It is interesting to note, however, that the J_H cluster is not CpG demethylated, despite bearing many other hallmarks of epigenetic activation described above. Fourth, μ_o and I_μ sterile transcripts initiate at PQ52 and $E\mu$ respectively. The other D_H gene segments are marked by low levels of anti-sense transcripts that are directed away from $C\mu$ (Bolland et al. 2007; Chakraborty et al. 2007). Bolland et al. have proposed that these anti-sense transcripts initiate at $E\mu$, and proceed through the J_H region and into the DSP and DFL16.1 gene segments.

2.4 Chromosome Conformation of the D_H–$C\mu$ Domain

Chromosome conformation capture (3C and 4C) studies (Zhao et al. 2006; Sexton et al. 2009; Vassetzky et al. 2009) of the pre-rearrangement IgH locus reveal interactions of $E\mu$ with sequences $5'$ of DFL16.1 ($5'$DFL) and with the $3'$ regulatory region ($3'$RR) that is centered approximately 200 kb from $E\mu$ (C. Guo and R. Sen, unpublished observations). The $3'$RR (see Fig. 1) consists of several DHSs (HS 1, 2, 3a, 3b, 4–7) spread over approximately 35 kb (Garrett et al. 2005). The cluster comprising HS 1, 2, 3a, and 3b lies within an inverted repeat region extending 25 kb beyond the exons encoding $C\alpha$. These sites, which include the $3'$ $E\alpha$ enhancer (Dariavach et al. 1991; Michaelson et al. 1995), are found only in activated B cells (Giannini et al. 1993; Madisen and Groudine 1994), and likely regulate switch region transcription associated with class switch recombination (Manis et al. 2002; Wuerffel et al. 2007; Dunnick et al. 2009). Sites 4–7 are present in a pro-B cell line and therefore may be functional at early stages of B cell differentiation (Giannini et al. 1993; Michaelson et al. 1995; Saleque et al. 1997). Interestingly, CpG residues within HS 4–7 are partially demethylated in pro-B cells (Giambra et al. 2008), analogous to the demethylation seen at PQ52 and $E\mu$.

These interaction studies are consistent with a three-loop configuration for the $3'$ end of the IgH locus (Fig. 2a). The smallest loop of approximately 5 kb is generated by $E\mu$-PQ52 interaction. The size of this loop is too small to be convincingly determined by 3C studies, and its existence is inferred from the observation that the PQ52 DHS and μ_o transcripts initiated at PQ52 are $E\mu$-dependent. This loop coincides with the region of maximum epigenetic activation described above and contains DQ52 and the four J_H gene segments. It also corresponds closely to the region that binds the highest levels of RAG1 and RAG2 proteins within the germline IgH locus (Ji et al. 2010). Localization of RAG proteins to this domain may be mediated by the recognition of high levels of H3K4me3 by the PHD domain of RAG2 (Liu et al. 2007b; Matthews et al. 2007; Ramon-Maiques et al. 2007) as well as RAG1 recognition of the J_H-associated RSSs (Ji et al. 2010). In this regard, the particularly accessible chromatin structure of this region indicated by its excessive DNase I sensitivity may be an essential feature of RAG1 recruitment to this region. Thus, the epigenetic state of this small loop contributes in multiple ways to generate the RAG-rich recombination center.

The next biggest loop is generated by interaction of $5'$DFL sequences with $E\mu$. This 58 kb loop contains all D_H gene segments other than DQ52. Most of this loop

Epigenetic Features that Regulate IgH Locus Recombination and Expression

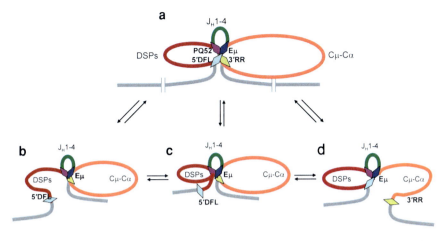

Fig. 2 Model for a dynamic three-loop configuration at the 3′ part of the IgH locus (**a**) three-loop configuration predicted from 3C and 4C studies of WT IgH alleles. The regulatory sequences Eμ (*blue*), PQ52 (*purple*), 3′RR (*yellow*) and the looping site 5′DFL (*light blue*) are shown as diamonds that are clustered in this configuration. The smallest PQ52-Eμ loop, shown in *green*, contains DQ52 and the four J_H gene segments. This region is maximally sensitive to nuclease digestion, is marked with highest levels of acetylated histones H3/H4 and H3K4me3 and coincides closely with the RAG1/2-rich recombinase center identified by Ji et al. (2010). The next biggest loop in red (labeled DSPs) contains all other D_H gene segments, with DFL16.1 (the 5′-most D_H gene) located close to the base of the loop near 5′DFL. Most of this loop is marked with heterochromatic H3K9me2 except at the bases (PQ52 and 5′DFL). The largest loop, of approximately 200 kb, forms between Eμ and the 3′RR; it contains the exons encoding all IgH constant regions. **b–d** Alternative conformations of the 3′ part of the IgH locus. The three-loop configuration shown in (**a**) is likely in dynamic equilibrium with a variety of two-loop configurations such as those shown in (**b**) and (**d**). **c** represents a situation where one of the DSP-associated promoters transiently interacts with Eμ, thereby bringing a DSP gene segment to the base of the loop and close to the RAG-rich recombination center to allow DSP to J_H rearrangements

is marked by heterochromatin-associated H3K9me2 modification, is insensitive to DNase I digestion and contains low levels of anti-sense transcripts. Only DFL16.1 lies in a pocket of active chromatin that corresponds to the base of the loop. The spatial proximity of DFL16.1 to Eμ in this configuration suggests that Eμ may have a role in inducing activation modifications at this location. Consistent with this view, Eμ-deleted IgH alleles lack the peak of histone acetylation at DFL16.1 (Chakraborty et al. 2009).

Finally, the largest loop is generated by interaction of Eμ with sequences located within HS5-7 of the 3′RR. This part of the 3′RR has been previously shown to contain hyperacetylated histones H3 and H4 in pro-B cell lines (Giambra et al. 2008) and its physical proximity to Eμ suggests that Eμ may contribute to histone acetylation in this region as well. Preliminary studies with Eμ-deficient pro-B cell lines indicate that activation modifications within HS5-7 are partially Eμ-dependent (C. Guo and R. Sen, unpublished observations). Although histone modification status of most of this 200 kb loop has not been systematically examined, H3 and H4 acetylation reduces to background levels within 6–8 kb 3′ of Eμ and heterochromatic H3K9me2 is present at Cγ3 exons (located 61 kb 3′ of Eμ) and at Cα exons (located 162 kb 3′ of Eμ). Extrapolating from these albeit restricted regions, we surmise that the rest of this loop is also likely to be in a relatively inactive chromatin state in pro-B cells. Thus, other than at the RAG-rich recombination center, the only pockets of epigenetically active chromatin in the germline IgH locus are found at the bases of the two larger loops. Interestingly, these are also the only regions that contain hypomethylated CpG DNA.

We note that biochemical chromosome conformation assays measure an ensemble average of locus conformations. Therefore, the three-loop configuration envisaged for this part of the IgH locus is unlikely to be a stable structure, or one that exists in all pro-B cells. Rather the proposed loops represent sites of most probable interactions and we imagine a dynamic state within cells where these loops form and break continuously (Fig. 2b–d). Whilst the three-loop configuration may exist in some cells within a population, or for some time within a cell, it will soon be replaced by various other single- or double-looped transient configurations.

2.5 Functional Implications of the Structure of the 3′ IgH Domain

The most prominent feature of the proposed structure is that most D_H gene segments (other than DQ52) are located within a domain that is tethered by Eμ. We have previously proposed that the pocket of activated chromatin around DFL16.1 may be responsible for its higher utilization in D_H rearrangements compared to the DSP gene segments, for example, by increasing RAG1/2 binding to active chromatin around this gene segment. The current observation that places DFL16.1 in close physical proximity to Eμ leads to a more satisfactory explanation for its greater usage. Since Eμ is located close to the RAG-rich J_H region, proximity of DFL16.1 to Eμ brings DFL16.1 close to the RAG-rich domain. In this configuration chromatin modifying enzymes associated with Eμ induce epigenetic alterations around DFL16.1 and RAG proteins tethered near the J_HS can recognize these alterations to initiate V(D)J recombination. The 'symmetrical' disposition of DQ52 and DFL16.1 to the RAG-rich domain brought about by the 5′DFL-Eμ looping may therefore provide the basis for over-utilization of these two gene segments in D_H to J_H recombination (Feeney 1990; Tsukada et al. 1990; Bangs et al. 1991; Chang et al. 1992; Atkinson et al. 1994; Nitschke et al. 2001).

The proposed structure explains the lower frequency of DSP gene rearrangements compared to either DFL16.1 or DQ52. We suggest that this is because DSP gene segments are located further away from the recombination center within the prominent DFL-Eμ loop. However, it also raises the question as to how RAG1/2 proteins bound at the recombination center ever find DSP gene segments to initiate rearrangements. Our earlier studies demonstrate that promoters associated with

Epigenetic Features that Regulate IgH Locus Recombination and Expression   47

DSP gene segments are activated for transcription after rearrangement, presumably by interacting with Eμ. That is, they are Eμ-dependent promoters like PQ52. We imagine that as the three major loops break and reform, it is possible that every so often Eμ associates with an unrearranged DSP promoter rather than with DFL16.1 (Fig. 2c). This would transiently bring that DSP gene segment and its associated RSS into the vicinity of the RAG-rich J$_H$ region, thereby permitting DSP to J$_H$ recombination to proceed (Fig. 2). Since the most frequent Eμ loops occur to DFL16.1, DSP rearrangements occur less frequently than DFL16.1 rearrangements.

The proposed structure also allows a plausible explanation for the observation that a V_H gene segment introduced very close to DFL16.1 appears to recombine preferentially with DQ52 (located 50 kb away) rather than with germline D_H gene segments located closer (Bates et al. 2007). The location of the knocked-in V_H gene places it at the base of the 5′DFL-Eμ loop and therefore close to the DQ52-J$_H$ containing RAG-rich domain. Normally, RAG-initiated cleavage at J$_H$-associated RSSs finds only DQ52- or DFL16.1-associated RSS within this chromatin domain, thereby leading to primarily DQ52- or DFL16.1 to J$_H$ recombination. Even when breaks occur at the DQ52-RSS, they would most readily find a complementary J$_H$-RSS for synapsis. However, with a V_H-RSS in spatial proximity to DQ52 due to 5′DFL-Eμ interaction, RAG-initiated cleavage at DQ52 would find a complementary RSS associated with the knocked-in V_H gene. Conversely, breaks at the knocked-in V_H-RSS induced by its spatial proximity to the RAG-rich domain could synapse with the DQ52-RSS to promote V_H to DQ52 rearrangement. This hypothesis implies that the DFL16.1-RSS becomes relatively RAG-poor compared to the DQ52-RSS because it is now located 5 kb away from the base of the loop.

A key feature of our model is that recombination to unrearranged D_H gene segments initiates within the RAG-rich DQ52-J$_H$ domain and proceeds within the chromatin limits of the 5′DFL-Eμ loop. This leads to the prediction that if the 5′DFL to Eμ loop could be expanded to include V_H gene segments, then V_H to germline D_H rearrangements, particularly to DQ52, may be possible. For example, in the absence of the 5′DFL looping site, Eμ would loop to some site in the V_H region, thereby bringing this region into spatial proximity of the RAG-rich domain. Now RAG-induced initial cleavage at a DQ52 RSS could potentially find a complementary RSS to synapse with not only at one of the J$_H$S (leading to canonical DJ$_H$ rearrangement), but also near a spatially proximal V_H gene segment resulting in V_H to DQ52 rearrangements. As discussed above, the proclivity of the nearest V_H gene segment for rearrangement would be affected by the distance of its associated RSS from the base of the loop. Conversely, the propensity of DFL16.1 rearrangements would be reduced because this gene segment would no longer be located near the base of a Eμ-tethered loop.

This may explain in part the recent observations of Giallourakis et al. (2010), who generated a large deletion in the IgH locus that spans the region from just 5′ of V_H81X till just after DFL16.1. Such a deletion removes the 5′DFL looping site and places the most proximal V_H gene segment about 9 kb 5′ of the nearest DSP gene segment. Whereas V_H recombination is normally restricted to the B lineage, they found that V_H7183 genes at the modified IgH locus rearranged in thymocytes. Breakdown of the strict regulation of V_H recombination led Giallourakis et al. to propose the presence of an important regulatory element in the deleted region. The 5'DFL looping site is a likely candidate for being this putative regulatory element. In the absence of this element, Eμ would loop to the next available upstream site which, based on the recombination results, probably lies within the proximal V_H7183 gene family. In this configuration, one or more V_H gene segments would be incorporated into this loop and those that lie close to the base of the loop will be in spatial proximity to the RAG-rich recombination center. In other words, the spatial location of one or more V_H gene segments would be analogous to the position of D_H gene segments on wild-type alleles. Since D_H genes are known to rearrange in thymocytes, analogously positioned V_H gene segments may become similarly permissive for recombination.

We have previously proposed that D_H rearrangement by deletion may have been selected as the preferred mode of DJ_H recombination because deletion brings V_H genes closer to the D_H–Cμ domain, and thereby possibly under the influence of Eμ (Chowdhury and Sen 2004). The looped configuration of the D_H–Cμ region provides another perspective on the lack of inversional D_H to J_H recombination. In the case of D_H recombination by deletion, the PQ52 DHS is lost (except when DQ52 recombines) and a different D_H-associated promoter is brought near J_H. Promoter activation by proximity to Eμ leads to transcription and increased nuclease accessibility at the DJ_H junction (see below). The resulting configuration resembles the germline situation with Eμ interacting with two regulatory sequences: the promoter associated with the rearranged D_H gene segment and 5'DFL. On alleles that rearrange D_H with inversion, the locus would retain PQ52 at a different position relative to Eμ and DFL16.1. Such a configuration would contain three regulatory sequences (the rearranged D_H promoter, PQ52 and 5'DFL) that could potentially 'sequester' Eμ activity. Such a chromatin configuration could be detrimental to V_H recombination, for example by reducing Eμ–dependent activation of the DJ_H junction for RAG-induced cutting or diffusing the RAG-rich recombination center over a larger portion of the genome.

3 The Second Step of IgH Gene Assembly: V_H to DJ_H Recombination

The second step of IgH gene assembly leads to the recombination of a V_H gene segment to the DJ_H junction. V_H recombination proceeds by deletion and the shortest recombinational event deletes approximately 100 kb of intervening DNA (when the 3' most V_H gene, V_H81X, recombines to the 5'-most D_H segment, DFL16.1). Utilization of distal V_HJ558 or V_H3609 families requires juxtaposition of gene segments located more than 2 Mb away from the DJ_H junction. Thus, it is

Epigenetic Features that Regulate IgH Locus Recombination and Expression

quite remarkable that V_H recombination occurs precisely to the RSS associated with the DJ_H junction and not to RSSs associated with unrearranged D_Hs located 5′ of the DJ_H junction. Specificity of V_H recombination is further highlighted by the fact that the closest germline D_H gene segment is only 4 kb 5′ of the DJ_H junction. The basis for this specificity may lie in several epigenetic features that distinguish DJ_H junctions from the multiple unrearranged D_H gene segments that lie 5′ to it.

3.1 Chromatin Structural Changes that Accompany D_H to J_H Recombination

D_H to J_H rearrangement activates transcription from the promoter located 5′ of the recombined D_H gene segment. This is evident from increased transcription as well as increased RNA polymerase recruitment selectively to DJ_H junctions (Chakraborty et al. 2007). DJ_H junctions present on $E\mu$-deleted IgH alleles are transcriptionally inactive indicating that rearranged D_H promoters are activated by $E\mu$ (unpublished observations). Interestingly, D_H-associated promoters appear to be bi-directional, leading to increased transcription towards and away from the $C\mu$ exons. However, levels of the anti-sense oriented transcript (directed away from $C\mu$) decrease rapidly with increasing distance from the promoter, perhaps due to de-stabilization of this transcript by lack of splicing or ineffective transition of RNA $PolII$ to its elongating form.

The simplest way to explain activation of DJ_H promoters, but not upstream germline D_H promoters, by $E\mu$ is that $E\mu$ activates transcription effectively only from the most proximal promoter. In the unrearranged IgH locus, the problem of tandem promoters and selective activation of the most $E\mu$-proximal promoter is less acute. Prior to D_H rearrangement the promoter most proximal to $E\mu$ is PQ52 and this interaction results in μ_o transcripts. Because of the gap between DQ52 and the next D_H gene segment, the next closest D_H-associated promoter is approximately 21 kb from $E\mu$. Perhaps $E\mu$ cannot activate this promoter because of the distance, or because of sequestration of $E\mu$ by PQ52 and 5′DFL sequences, or because of the heterochromatic nature of the intervening region between PQ52 and the upstream D_H gene segments. After D_H rearrangement, however, PQ52 is deleted (except in the special case of DQ52 rearrangement) and the promoter associated with the rearranged D_H gene segment takes its place. Now the closest upstream promoter is less than 5 kb from the rearranged DJ_H promoter. Yet, $E\mu$ activity is restricted to the DJ_H promoter, indicating that the upstream germline D_H promoter cannot effectively compete for $E\mu$.

Several epigenetic features distinguish DJ_H junctions from unrearranged D_H gene segments. First, recombined D_H gene segments lose repressive H3K9me2 modifications that are present at germline DSP and DST4 gene segments. Loss of H3K9me2 is accompanied by gain of hyperacetylated histones H3 and H4, moderate levels of H3K4me2 and high levels of transcription-associated H3K4me3 (R. Subrahmanyam and R. Sen, unpublished observations).

Second, DJ_H junctional DNA is selectively (CpG) hypomethylated (R. Selimyan and R. Sen, unpublished observations). Third, the region surrounding DJ_H junctions becomes more sensitive to DNase I digestion than the same D_H gene segment in germline configuration. The most remarkable feature of all these changes in chromatin structure is that they are restricted to the DJ_H junction and do not exceed more than a few kb 5′ (R. Subrahmanyam and R. Sen, unpublished observations). Our working hypothesis is that this confluence of epigenetic activation at DJ_H junctions facilitates V_H recombination to the DJ_H-associated RSS, while excluding the more numerous RSSs associated with unrearranged D_H gene segments. Consistent with this idea, RAG recruitment to DJ_H rearranged alleles is also focused near the DJ_H junction and does not extend to unrearranged D_H gene segments located within a few kb 5′ (R. Subrahmanyam and R. Sen, unpublished observations). The close correlation between RAG recruitment and epigenetic state of DJ_H junctions further strengthens the idea that chromatin structure dictates where the highest levels of RAG proteins are bound.

The high levels of activating modifications at DJ_H junctions can be understood quite simply. Prior to D_H rearrangements J_H regions have the highest levels of nuclease accessibility and activation-associated histone modifications. Thus, D_H recombination brings a D_H gene segment into this active domain. Since the chromatin state of the J_H region is $E\mu$-dependent (Chakraborty et al. 2009), the recombined D_H gene segment comes under the influence of $E\mu$. It is more difficult to explain why the influence of $E\mu$ drops off so rapidly 5′ of the DJ_H junction. One possibility is that chromatin modifying enzymes, such as histone acetyltransferases, that are delivered by $E\mu$ are not sufficiently processive in order to migrate further 5′. It must also be recalled that the region immediately 5′ of the recombined D_H has several features of heterochromatin in the germline configuration. If this state accompanies the rearranging D_H gene segment, then for active chromatin marks to spread into the 5′ DJ_H region these negative histone modifications must first be removed. Therefore, it follows that if H3K9me2 demethylases are not concurrently recruited to the upstream D_H gene segments, the heterochromatic state would persist. Furthermore, we have previously shown that maintenance of H3K9me2 in the DSP repeats requires continuous deacetylation (Chakraborty et al. 2007). This raises the possibility that histone deacetylases may be brought in with the rearranged gene segment actively precluding the spread of active chromatin into adjacent germline D_H gene segments. Finally, it is possible that elevated antisense transcription just 5′ of the DJ_H junction may play a role in preventing the spread of active chromatin. Our working model is that the distinctive chromatin state of DJ_H junctions compared to upstream unrearranged D_H gene segments plays an important role in targeting V_H recombination to the DJ_H-RSS. Thus, unraveling the mechanisms that generate this boundary are essential to fully understand the specificity of V(D)J recombination. Future studies of the location and function of histone deacetylases or demethylases that maintain the boundary between the active DJ_H junction and inactive germline D_Hs will likely be informative in this regard.

3.2 How does DJ$_H$ Recombination Activate V$_H$ Recombination?

The chromatin structure of germline and DJ$_H$ recombined alleles provides a plausible model for secondary V$_H$ recombination. The close correspondence between active histone modifications, nuclease sensitivity, and the location of RAG proteins on unrearranged and DJ$_H$-rearranged alleles suggests that RAG proteins are recruited to these sites as a consequence of chromatin structure. To initiate V(D)J recombination, RAG proteins bind to local J$_H$-associated RSSs and scan the "environment" for a complementary RSS to synapse with. The looped configuration of the D$_H$-Eμ region discussed in the preceding sections provides proximity to D$_H$-associated RSSs and, in particular, DFL16.1- and DQ52-associated RSSs which are spatially close to the RAG-rich J$_H$ region. Once J$_H$-associated RAG proteins 'find' a D$_H$ RSS, synapsis occurs and D$_H$ to J$_H$ recombination proceeds. After D$_H$ rearrangement, the 5'-D$_H$ RSS is in the RAG-rich domain together with RSSs associated with remaining unrearranged J$_H$ gene segments. If J$_H$ RSS-associated RAG1/2 initiate a new round of rearrangements, the outcome is secondary D$_H$ recombination. However, if RAG1/2 bound to the DJ$_H$-associated RSS initiate recombination, then productive synapsis can occur only with a V$_H$ RSS. Once such a partner is found, V$_H$ to DJ$_H$ recombination can proceed. In this model V$_H$ recombination occurs after D$_H$ recombination because only then is the 5'-RSS of a D_H gene segment located within the recombination center to initiate recombination. The situation with DQ52 remains murky. Both DQ52 RSSs are located within the recombination center in the germline state. Hence, why does RAG binding to the 5' DQ52 RSS not initiate V$_H$ to DQ52 rearrangements? Our working hypothesis is that this is due to the local availability of J$_H$-associated complementary RSSs and the relative unavailability of the V$_H$-associated complementary RSSs. As noted in our interpretation of the Bates et al. and Giallourakis et al. studies in the preceding section, such rearrangements do indeed proceed when a V$_H$ region is artificially placed in spatial proximity to DQ52. The model proposed above does not require secondary activation of V$_H$ gene accessibility, for example by additional regulatory sequences that are activated after DJ$_H$ recombination. Rather, V$_H$ genes may be fully accessible prior to initiation of rearrangements, but recombine only after D$_H$-RSSs are brought into the RAG rich domain by D$_H$ to J$_H$ recombination. Early accessibility of V$_H$ genes is supported by the observation that sterile V$_H$ transcription is easily apparent on germline IgH alleles (Yancopoulos and Alt 1985).

3.3 Regulation of V$_H$ Gene Recombination

The genomic region that encodes V$_H$ gene segments spans 2.5 Mb in the mouse (Johnston et al. 2006). Within this are located 110 functional V$_H$ gene segments and 85 pseudo-V$_H$ gene segments. Based on sequence similarity V$_H$ genes can be divided into 16 families. The V_HJ558 gene family contains the largest numbers of functional gene segments and is located in the 5′ 1.6 Mb of the V_H locus. Members of the V_H3609 family are interspersed with V_HJ558 genes within this region. The 3′-most V_H gene family comprises members of the V_H7183 family, of which 7183.2.3 (also known as V_H81x) is the 3′-most functional V_H gene segment. The V_H7183 family extends approximately 300 kb and is interspersed with genes of the V_HQ52 family. The remaining 700 kb of the locus between J558/3609 and 7183/Q52 families contains 12 other gene families including V_HS107, V_HSM7, V_H36-60, V_H11, $V_HVGAM3.8$, V_HJ606 and V_H10. Each of these V_H gene families contains only 1–6 members. V_H7183 and V_HQ52 families, which together consist of approximately 19 gene segments, are often referred to as proximal V_H genes because they are located closest to the D_H–$C\mu$ part of the locus. Conversely, V_HJ558 and V_H3609 families, consisting of 60 gene segments, are referred to as distal V_H genes.

V_H gene recombination is regulated in three important ways. First, V_H recombination follows D_H recombination during B cell development. A plausible mechanism for this was discussed in the preceding section. Second, several lines of evidence demonstrate that recombination of proximal and distal V_H gene segments is regulated very differently. This aspect of V_H gene regulation is discussed below. Thirdly, V_H recombination is subject to feedback inhibition, which accounts for allelic exclusion at the IgH locus. Allelic exclusion and the mechanisms that mediate it have been discussed in recent reviews (Cedar and Bergman 2008; Vettermann and Schlissel 2010; Hewitt et al. 2010) and will not be discussed here.

3.4 Differential Regulation of Proximal and Distal V_H Recombination

Recognition that proximal and distal V_H genes are differentially regulated came from the observation that proximal V_H genes recombined much more frequently than distal V_H genes during fetal B cell ontogeny (Yancopoulos et al. 1984; Perlmutter et al. 1985; Lawler et al. 1987; Jeong and Teale 1988; Malynn et al. 1990; ten Boekel et al. 1997). Based on current information, the simplest explanation for this is that IL-7/IL-7R signaling pathway, which is important for distal V_H recombination, is not a major player in fetal B lymphopoiesis (Carvalho et al. 2001; Miller et al. 2002). Instead, signaling via thymic stromal lymphopoietin (TSLP) and Flt3 ligand appears to dominate in the fetus (Vosshenrich et al. 2003, 2004; Jensen et al. 2007, 2008), suggesting that these cytokines activate proximal V_H recombination. In the adult bone marrow IL-7 and Flt3L work together to activate the entire V_H repertoire for recombination. Consequently, double-deletion of IL-7 and Flt3L results in a complete block of B cell development (Jensen et al. 2008). The mechanism by which TSLP and Flt3L activate proximal V_H gene recombination remains to be determined.

The importance of IL-7/IL-7R signaling for distal V_H gene recombination is evident not only from the analysis of IL-7- or IL-7R-deficient mice (Corcoran et al. 1998; Carvalho et al. 2001), but also from the analysis of STAT5A/B-double-deficient mice (Bertolino et al. 2005) and from the partial rescue of B lymphopoiesis in IL-7R-deficient mice by transgenic expression of a constitutively active form of STAT5b (Goetz et al. 2004). The IL-7/IL-7R pathway works by altering the chromatin state and transcriptional status (a measure of accessibility) of distal V_H genes (Chowdhury and Sen 2001; Johnson et al. 2004; Stanton and Brodeur 2005). This has been proposed to be mediated by recruitment of STAT5 to germline V_H gene promoters via interaction with DNA-bound octamer binding proteins (Bertolino et al. 2005). However, it should be noted that the substantial evidence in favor of a direct effect of IL-7R signaling on distal V_H recombination has been contested by a recent study. Malin et al. (2010) found that the B cell developmental defect caused by conditional deletion of STAT5 could be rescued by transgenic expression of Bcl2. In this study, distal V_H genes recombined efficiently in the absence of IL-7R signaling leading the authors to conclude that IL-7/IL-7R interaction provides cell viability rather than developmental signals. The possible sources of the apparent discrepancy between this and earlier studies have been addressed by Skok and colleagues (Hewitt et al. 2010).

Additionally, three transcription factor deficiencies selectively affect distal V_H recombination. The best characterized of these is Pax5, a B lineage-selective paired domain transcription factor whose germline deletion blocks B cell development at the pro-B cell stage (Urbanek et al. 1994). Cells that accumulate in the bone marrow of Pax5-deficient mice have normal D_H to J_H recombination (Nutt et al. 1997) and mostly proximal V_H to DJ_H recombination (Hesslein et al. 2003). Pax5 binding sites have been identified in a subset of mouse and human V_H gene segments (Zhang et al. 2006), where this transcription factor may directly recruit RAG proteins by protein/protein interactions. Recently, Busslinger and colleagues have shown that Pax5 binds close to multiple members of the V_H3609 gene family in the distal part of the murine V_H locus (Ebert et al. 2011). Many of these locations also bind the transcription factors E2A and CTCF, thereby forming a composite element of approximately 470 bp. This element has been named Pax5-associated intergenic repeat (PAIR) and 14 PAIR elements with varying degrees of similarity have been identified. Interestingly, PAIR sequences serve as promoters in transfection assays and may be a source of anti-sense oriented transcripts in the distal V_H locus. Their location within the IgH locus and association with Pax5 has led to the proposal that they may mediate Pax5-dependent chromosome compaction and distal V_H gene recombination (Ebert et al. 2011).

YY1 is a ubiquitously expressed zinc finger transcription factor with homology to the *Drosophila melanogastor* gene, *polycomb* (Calame and Atchison 2007). B lineage-specific deletion of YYl results in a developmental block similar to that of Pax5-deficiency. In pro-B cells from these mice D_H and proximal V_H gene recombination are unaffected, but distal V_H recombination is substantially reduced

(Liu et al. 2007a). The third transcription factor that regulates V_H recombination is Ikaros; however, the situation with Ikaros deficiency is more complex. In the absence of Ikaros, B cell development is blocked at the pro-B cell stage with hardly any proximal or distal V_H recombination (Reynaud et al. 2008). However, Ikaros deficiency also leads to reduced RAG expression. To determine whether the phenotype of Ikaros-deficient pro-B cells is entirely dependent on the lack of RAG protein, Reynaud et al. ectopically expressed RAG1 and RAG2 in Ikaros-deficient precursors. They observed only proximal V_H gene recombination in these RAG-reconstituted pro-B cells, suggesting that Ikaros deficiency selectively affects distal V_H recombination. Consistent with this interpretation, sterile transcription of distal V_H gene segments, but not proximal V_H gene segments, is attenuated in *Ikaros*-deficient pro-B cells. Finally, B lineage-specific deletion of the histone H3 lysine 27 methyltransferase Ezh2 also blocks distal V_H recombination in pro-B cells (Su et al. 2003).

The mechanism(s) by which Pax5, YY1, Ikaros and Ezh2 enhance distal V_H recombination is not known. It is interesting, however, that all of these proteins are involved in gene repression. Ikaros and Pax5 are context-specific gene activators or repressors. The repressive function of Ikaros is manifest in the context of the NuRD co-repressor complex (Kim et al. 1999; Sridharan and Smale 2007), whereas that of Pax5 is mediated via Groucho-related co-repressor complexes (Eberhard et al. 2000). YY1 and Ezh2 are mammalian orthologs of *polycomb* group genes that were characterized as transcriptional repressors in *Drosophila melanogastor* (Muller and Kassis 2006; Schwartz and Pirrotta 2007). Despite this apparently unifying characteristic, it is unclear whether these proteins participate directly or indirectly at the V_H locus, whether they act in the same, or parallel, pathways of V_H activation and how the proximal V_H genes are excluded from their influence.

While considering the mechanisms by which distal and proximal V_H genes are differentially regulated, it is important to note that it is not clear how far into the V_H7183/V_HQ52 region proximity effects extend. The V_H7183/V_HQ52 gene families occupy approximately 300 kb of genomic DNA $5'$ of DFL16.1 and contain 19 functional gene segments. In WT C57BL/6 strain, the utilization of these gene segments appears to be inversely proportional to the distance from the D_H–$C\mu$ part of the locus (Williams et al. 2001). That is, the most $3'$ V_H gene segments rearrange more frequently than those that lie at the $5'$ end of the 7183/Q52 family. Because degenerate primers are often used to assay V(D)J recombination in various genetic deficiencies, a positive signal could come from the closest 2–3 V_H gene segments, or from a more equal distribution of all the gene segments in these families. Indeed, the distribution of proximal V_H gene utilization could be different even within the several genotypes that are currently considered to not affect proximal V_H recombination. It is likely that accurate estimates of where proximity effects end will be important for future mechanistic understanding of the distinctive regulation of proximal V_H gene recombination.

3.5 Epigenetic Features that Could Mediate Differential V_H Gene Recombination

3.5.1 Spatial Configuration of Distal and Proximal V_H Genes

Using fluorescence in situ hybridization (FISH) with BAC probes that recognized different parts of the V_H locus, Kosak et al. (2002) were the first to demonstrate that the 5′ and 3′ V_H gene families are brought into spatial proximity in pro-B cell nuclei. They referred to the phenomenon as locus compaction and reasoned that such conformational changes facilitated recombination by bringing together distantly located gene segments. Using three-color FISH, Sayegh et al. (2005) provided evidence that compaction involved chromosome looping. In taking FISH analysis a step further these authors examined the spatial relationship between distal, or proximal, V_H gene families and the constant region part of the IgH locus. The correlation between locus compaction and V_H recombination was further extended by Busslinger and colleagues who found that the IgH locus did not undergo locus compaction in Pax5-deficient pro-B cells (Fuxa et al. 2004). Since the observation in $Pax5^{-/-}$ pro-B cells, lack of locus compaction has also been observed in Ikaros and YY1-deficient pro-B cells where distal V_H gene recombination is reduced (Liu et al. 2007a; Reynaud et al. 2008). Importantly, FISH analyses also showed that spacing between proximal V_H genes and the $C\mu$ part of the IgH locus was not affected in YY1- or Pax5-deficient pro-B cells. Thus, these studies demonstrate that spatial reconfiguration of proximal and distal V_H genes occur by different mechanisms.

The three-dimensional (3D) structure of the germline IgH locus has been elegantly inferred from a combination of 3D-FISH studies and trilateration analyses (Jhunjhunwala et al. 2008). By comparing E2A-deficient pre-pro-B cells with RAG2-deficient (but E2A-sufficient) pro-B cells, Jhunjhunwala et al. noted several important features of the IgH locus. In pre-pro-B cells, the spatial distance measurements made by FISH fit best with the computational major loop subcompartment (MLS) model. In this configuration, the IgH locus can be represented as a set of domains (sub-compartments) with each domain consisting of multiple chromosomal loops. The proximal V_H genes are already located close to the D_H/J_H gene segments in this configuration, but the distal V_H genes are not. The trilateration studies therefore also strengthen the idea that proximal and distal V_H compaction is mediated by different mechanisms. Upon further differentiation to pro-B cells, the orientation of the proximal V_H genes changes relative to the D_H/J_H gene segments and, most importantly, the distal V_H genes are brought into spatial proximity to the 3′ end of the IgH locus. This reconfiguration of distal V_H genes in pro-B cells likely represents the step of locus compaction that is abrogated in the absence of Pax5, YY1 and Ikaros. These observations also show that proximal and distal V_H compaction occur at different stages of differentiation. Assuming that re-orientation of proximal V_H genes that occurs during pre-pro-B to pro-B transition would not be detected by the earlier two-color FISH studies, we suggest that compaction between proximal V_Hs and the D_H–$C\mu$ part of the locus that is seen in Pax5- and YY1-deficient pro-B cells represents a pre-pro-B-like MLS structure. Overall, the close correlation between distal V_H compaction visualized by FISH and distal V_H recombination strongly argues that locus compaction is directly involved in V(D)J recombination.

3.5.2 Distinct Histone Modifications may Regulate Proximal and Distal V_H Recombination

The earliest evidence for differential epigenetic regulation of V_H gene recombination was the demonstration that IL-7/IL-7R signaling resulted in histone H3 and H4 hyperacetylation of distal but not proximal V_H genes in primary pro-B cells (Chowdhury and Sen 2001). It was inferred that the requirement for ex vivo treatment of pro-B cells with IL-7 was due to limiting cytokine concentrations in vivo. Incubation with IL-7 ex vivo provided uniform activation of most cells in culture and permitted biochemical changes to be detected. The striking observation that representative V_HJ558, but not V_H81X, gene segments became hyperacetylated suggested differential regulation of the distal and proximal parts of the V_H locus. An obvious caveat in these studies, which were carried out before the genomics era, was that the observations were based on a few gene segments and extrapolated to more general conclusions. For example, it was not clear whether IL-7-induced hyperacetylation was widespread amongst V_HJ558 family members or restricted to a few gene segments. Nor were the boundaries of the IL-7-inducible domain delineated in these studies. Nevertheless, these basic findings were reproduced in different contexts, including cell culture models (Stanton and Brodeur 2005), specific genetic knockouts (STAT5A/B) (Bertolino et al. 2005) and by FISH studies (Roldan et al. 2005), thereby substantiating a role for IL-7 in activating distal V_H gene segments.

Recently Feeney and colleagues provided a unique perspective on the differential regulation of proximal and distal V_H gene recombination. Taking a cue from the lack of distal V_H gene recombination in Ezh2$^{-/-}$ pro-B cells, they examined the distribution of H3K27me3 (the product of Ezh2 activity) across the V_H locus in pro-B cells. Strikingly, only the region encompassing the proximal V_H7183 gene family contained H3K27me3, whereas the distal V_HJ558 and V_H3609 gene families were reciprocally marked with activation-associated H3K36me3 (Xu et al. 2008). In Pax5-deficient pro-B cells and fetal liver pro-B cells, both of which undergo largely proximal V_H recombination, H3K27me3 levels were reduced at the V_H7183 genes. They concluded that inhibitory modifications of the proximal V_H genes may be essential for distal V_H recombination, perhaps by suppressing proximal V_H recombination. This is interesting because it raises the idea, for the first time, that the choice of distal V_H recombination may depend upon a 'competition' between mutually antagonistic epigenetic modifications of distal versus proximal V_H genes. Although the relationship between Pax5 and proximal V_H H3K27me3 is not clear, one possibility is that Pax5 binding sites in the V_H7183

Epigenetic Features that Regulate IgH Locus Recombination and Expression 57

family may recruit Ezh2 to this region. Why this does not occur in fetal liver pro-B cells remains to be determined. As noted before, identifying the limits of the H3K27me3 domain and the relative location of Pax5 binding sites will be useful in understanding the mechanisms involved.

3.6 Coordinating Locus Compaction and Histone Modification of V_H Genes

It is interesting to consider the relationship between chromosome conformation and histone modification state of the V_H genes. One key observation is that RAG-deficient pro-B cells already have the IgH locus in a compacted state, yet hyperacetylation and sterile transcription of distal genes requires additional IL-7 treatment. The simplest interpretation of this observation is that chromatin compaction and distal V_H gene segment hyperacetylation are independently regulated. Analysis of Pax5-deficient pro-B cells also leads to a similar conclusion. Hesslein et al. (2003) found that distal V_H genes were actively transcribed and marked with hyperacetylated histones in Pax5-deficient pro-B cell cultures. Because the IgH locus is de-contracted in Pax5-deficient cells, these observations indicate that locus compaction is not a pre-requisite for IL-7-induced histone hyperacetylation and transcription of V_H genes. That distal V_H recombination is impaired in both Pax5- and IL-7-deficiency therefore leads to the important conclusion that compaction and local chromatin changes are both essential for distal V_H recombination.

How could these two pathways synergistically activate distal V_H recombination? Our working hypothesis is that IL-7-induced chromatin structural changes generate accessibility that is necessary for DJ_H-bound RAG proteins to bind to complementary RSSs associated with V_H gene segments, while Pax5-induced compaction brings distal V_H gene segments close to the DJ_H part of the IgH locus. In other words, in the absence of Pax5, distal V_H gene segments may be structurally accessible to RAG recombinase, but they would be located too far from the RAG-rich recombination center near DJ_H junctions to effectively recombine. Conversely, distal V_H genes would be spatially close to the recombination center in the absence of IL-7, but in the absence of STAT5-dependent chromatin structural changes the RSSs would be inaccessible to synapse with DJ_H bound RAG proteins. Another way in which histone hyperacetylation may promote distal V_H recombination comes from the observation that the movement of one IgH allele to centromeric heterochromatin in activated mature B cells is prevented by co-stimulation with IL-7 (Roldan et al. 2005). Since distal V_H genes are inducibly hyperacetylated by IL-7 treatment of mature B cells (Chowdhury and Sen 2003), these observations suggest that histone acetylation prevents translocation of IgH alleles to centromeric heterochromatin. It is possible that even at the pro-B cell stage, IL-7-induced distal V_H acetylation reduces the tendency of distal V_H genes to translocate to centromeric heterochromatin and thereby maintain accessibility to RAG proteins.

Beyond its role in locus compaction, Pax5 has also been proposed to regulate the histone modification state of distal V_H genes. Johnson el al. found that distal V_H genes in Pax5-deficient pro-B cell cultures were marked with the repressive H3K9me2 mark (Johnson et al. 2004). They concluded that presence of H3K9me2 inhibited distal V_H recombination in Pax5-deficient pro-B cells as had been demonstrated with model recombination substrates (Osipovich et al. 2004). In view of the activation-associated modifications of distal V_H genes in Pax5-deficient pro-B cells described above, these observations are counterintuitive because they indicate that distal V_H genes contain both activating and repressive histone modifications in Pax5-deficient pro-B cells. This is reminiscent of bivalent chromatin observed in embryonic stem cells where transcription-associated H3K4me3 and repression-associated H3K27me3 modifications are simultaneously associated with a subset of gene promoters (Bernstein et al. 2006). The proposed recombination inhibitory effect of the repressive H3K9me2 mark at bivalently marked distal V_H gene segments also correlates with very low levels of transcription of bivalent ES cell genes. However, it remains possible that the apparently contradictory modifications noted in these studies may be due to different V_H gene segments being queried, different parts of V_H gene segments being queried or differences between the cell populations used in each study. Together with the Feeney study (Xu et al. 2008), these observations underscore the importance of repressive histone modifications in regulating V_H gene recombination. It is interesting that presence of Pax5 leads to reduced H3K9me2 at the distal V_H genes and, concurrently, increased H3K27me3 at the proximal genes. Additional studies will no doubt clarify the role of Pax5 in differentially modulating these two forms of negative modifications.

4 Conclusions

Work from many laboratories has made the IgH locus one of the best characterized developmentally regulated loci. Much is known about the several *cis*-elements that regulate IgH transcription and recombination, the various *trans*-acting factors that bind to these promoters and enhancers, and the broader role these factors play in B lymphocyte development. In recent years much of the activity in this area has focused on epigenetic mechanisms that participate in the control of IgH expression. In this review we summarize the salient epigenetic features that help to explain some of the accumulated observations associated with IgH expression. Concurrently, we identify aspects that remain murky and sometimes offer the possible mechanisms whose validity awaits to be established by experimentation. A case in point is the control of D_H gene rearrangements. The known epigenetic features of the pre-rearrangement D_H-$C\mu$ part of the locus offers insights into the over-representation of the two flanking D_H gene segments in VDJ recombined alleles. At the same time the available epigenetic and genetic information does not offer a ready explanation of how intervening D_H gene segments ever recombine, or why the majority of D_H rearrangements occur by deletion, rather than inversion, of the intervening DNA. Thus, for the latter we propose possible mechanisms that we expect will stimulate experimentation and, no doubt, criticism. Other currently unsolved questions we discuss are the secondary onset of V_H recombination, the precision of V_H recombination to the rearranged DJ_H junction and the differential regulation of distal versus proximal V_H gene recombination. While these questions may be considered by some in the community to be "details", we are optimistic that chromosomal chemistry of the IgH locus stands at the threshold of true mechanistic understanding.

Acknowledgments The authors are supported by the Intramural Research Program of the National Institute on Aging, Baltimore, MD.

References

Abarrategui I, Krangel MS (2009) Germline transcription: a key regulator of accessibility and recombination. Adv Exp Med Biol 650:93–102

Atkinson MJ, Chang Y, Celler JW, Huang C, Paige CJ, Wu GE (1994) Overusage of mouse DH gene segment, DFL16.1, is strain-dependent and determined by cis-acting elements. Dev Immunol 3:283–295

Bangs LA, Sanz IE, Teale JM (1991) Comparison of D, JH, and junctional diversity in the fetal, adult, and aged B cell repertoires. J Immunol 146:1996–2004

Bates JG, Cado D, Nolla H, Schlissel MS (2007) Chromosomal position of a VH gene segment determines its activation and inactivation as a substrate for V(D)J recombination. J Exp Med 204:3247–3256

Bernstein BE, Mikkelsen TS, Xie X, Kamal M, Huebert DJ, Cuff J, Fry B, Meissner A, Wernig M, Plath K, Jaenisch R, Wagschal A, Feil R, Schreiber SL, Lander ES (2006) A bivalent chromatin structure marks key developmental genes in embryonic stem cells. Cell 125:315–326

Bertolino E, Reddy K, Medina KL, Parganas E, Ihle J, Singh H (2005) Regulation of interleukin 7-dependent immunoglobulin heavy-chain variable gene rearrangements by transcription factor STAT5. Nat Immunol 6:836–843

Blackwell TK, Moore MW, Yancopoulos GD, Suh H, Lutzker S, Selsing E, Alt FW (1986) Recombination between immunoglobulin variable region gene segments is enhanced by transcription. Nature 324:585–589

Bolland DJ, Wood AL, Afshar R, Featherstone K, Oltz EM, Corcoran AE (2007) Antisense intergenic transcription precedes Igh D-to-J recombination and is controlled by the intronic enhancer Emu. Mol Cell Biol 27:5523–5533

Calame K, Atchison M (2007) YY1 helps to bring loose ends together. Genes Dev 21:1145–1152

Carvalho TL, Mota-Santos T, Cumano A, Demengeot J, Vieira P (2001) Arrested B lymphopoiesis and persistence of activated B cells in adult interleukin 7(-/-) mice. J Exp Med 194:1141–1150

Cedar H, Bergman Y (2008) Choreography of Ig allelic exclusion. Curr Opin Immunol 20: 308–317

Chakraborty T, Chowdhury D, Keyes A, Jani A, Subrahmanyam R, Ivanova I, Sen R (2007) Repeat organization and epigenetic regulation of the DH-Cmu domain of the immunoglobulin heavy-chain gene locus. Mol Cell 27:842–850

Chakraborty T, Perlot T, Subrahmanyam R, Jani A, Goff PH, Zhang Y, Ivanova I, Alt FW, Sen R (2009) A 220-nucleotide deletion of the intronic enhancer reveals an epigenetic hierarchy in immunoglobulin heavy chain locus activation. J Exp Med 206:1019–1027

Chang Y, Paige CJ, Wu GE (1992) Enumeration and characterization of DJH structures in mouse fetal liver. EMBO J 11:1891–1899

Chowdhury D, Sen R (2001) Stepwise activation of the immunoglobulin mu heavy chain gene locus. Embo J 20:6394–6403

Chowdhury D, Sen R (2003) Transient IL-7/IL-7R signaling provides a mechanism for feedback inhibition of immunoglobulin heavy chain gene rearrangements. Immunity 18:229–241

Chowdhury D, Sen R (2004) Regulation of immunoglobulin heavy-chain gene rearrangements. Immunol Rev 200:182–196

Cobb RM, Oestreich KJ, Osipovich OA, Oltz EM (2006) Accessibility control of V(D)J recombination. Adv Immunol 91:45–109

Corcoran AE, Riddell A, Krooshoop D, Venkitaraman AR (1998) Impaired immunoglobulin gene rearrangement in mice lacking the IL-7 receptor. Nature 391:904–907

Dariavach P, Williams GT, Campbell K, Pettersson S, Neuberger MS (1991) The mouse IgH 3'-enhancer. Eur J Immunol 21:1499–1504

Dunnick WA, Collins JT, Shi J, Westfield G, Fontaine C, Hakimpour P, Papavasiliou FN (2009) Switch recombination and somatic hypermutation are controlled by the heavy chain 3' enhancer region. J Exp Med 206:2613–2623

Eberhard D, Jimenez G, Heavey B, Busslinger M (2000) Transcriptional repression by Pax5 (BSAP) through interaction with corepressors of the Groucho family. Embo J 19:2292–2303

Ebert A, McManus S, Tagoh H, Medvedovic J, Salvagiotto G, Novatchkova M, Tamir I, Sommer A, Jaritz M, Busslinger M (2011) The distal V(H) gene cluster of the Igh locus contains distinct regulatory elements with Pax5 transcription factor-dependent activity in pro-B cells. Immunity 34:175–187

Featherstone K, Wood AL, Bowen AJ, Corcoran AE (2010) The mouse immunoglobulin heavy chain V-D intergenic sequence contains insulators that may regulate ordered V(D)J recombination. J Biol Chem 285:9327–9338

Feeney AJ (1990) Lack of N regions in fetal and neonatal mouse immunoglobulin V-D-J junctional sequences. J Exp Med 172:1377–1390

Ferrier P, Covey LR, Suh H, Winoto A, Hood L, Alt FW (1989) T cell receptor DJ but not VDJ rearrangement within a recombination substrate introduced into a pre-B cell line. Int Immunol 1:66–74

Fuxa M, Skok J, Souabni A, Salvagiotto G, Roldan E, Busslinger M (2004) Pax5 induces V-to-DJ rearrangements and locus contraction of the immunoglobulin heavy-chain gene. Genes Dev 18:411–422

Garrett FE, Emelyanov AV, Sepulveda MA, Flanagan P, Volpi S, Li F, Loukinov D, Eckhardt LA, Lobanenkov VV, Birshtein BK (2005) Chromatin architecture near a potential 3' end of the igh locus involves modular regulation of histone modifications during B-Cell development and in vivo occupancy at CTCF sites. Mol Cell Biol 25:1511–1525

Gauss GH, Lieber MR (1992) The basis for the mechanistic bias for deletional over inversional V(D)J recombination. Genes Dev 6:1553–1561

Giallourakis CC, Franklin A, Guo C, Cheng HL, Yoon HS, Gallagher M, Perlot T, Andzelm M, Murphy AJ, Macdonald LE, Yancopoulos GD, Alt FW (2010) Elements between the IgH variable (V) and diversity (D) clusters influence antisense transcription and lineage-specific V(D)J recombination. Proc Natl Acad Sci U S A 107:22207–22212

Giambra V, Volpi S, Emelyanov AV, Pflugh D, Bothwell AL, Norio P, Fan Y, Ju Z, Skoultchi AI, Hardy RR, Frezza D, Birshtein BK (2008) Pax5 and linker histone H1 coordinate DNA methylation and histone modifications in the 3' regulatory region of the immunoglobulin heavy chain locus. Mol Cell Biol 28:6123–6133

Giannini SL, Singh M, Calvo CF, Ding G, Birshtein BK (1993) DNA regions flanking the mouse Ig 3' alpha enhancer are differentially methylated and DNAase I hypersensitive during B cell differentiation. J Immunol 150:1772–1780

Goetz CA, Harmon IR, O'Neil JJ, Burchill MA, Farrar MA (2004) STAT5 activation underlies IL7 receptor-dependent B cell development. J Immunol 172:4770–4778

Hesslein DG, Pflugh DL, Chowdhury D, Bothwell AL, Sen R, Schatz DG (2003) Pax5 is required for recombination of transcribed, acetylated, 5' IgH V gene segments. Genes Dev 17:37–42

Epigenetic Features that Regulate IgH Locus Recombination and Expression   61

Hewitt SL, Chaumeil J, Skok JA (2010) Chromosome dynamics and the regulation of V(D)J recombination. Immunol Rev 237:43–54

Jensen CT, Kharazi S, Boiers C, Liuba K, Jacobsen SE (2007) TSLP-mediated fetal B lymphopoiesis? Nat Immunol 8:897

Jensen CT, Kharazi S, Boiers C, Cheng M, Lubking A, Sitnicka E, Jacobsen SE (2008) FLT3 ligand and not TSLP is the key regulator of IL-7-independent B-1 and B-2 B lymphopoiesis. Blood 112:2297–2304

Jeong HD, Teale JM (1988) Comparison of the fetal and adult functional B cell repertoires by analysis of VH gene family expression. J Exp Med 168:589–603

Jhunjhunwala S, van Zelm MC, Peak MM, Cutchin S, Riblet R, van Dongen JJ, Grosveld FG, Knoch TA, Murre C (2008) The 3D structure of the immunoglobulin heavy-chain locus: implications for long-range genomic interactions. Cell 133:265–279

Ji Y, Resch W, Corbett E, Yamane A, Casellas R, Schatz DG (2010) The in vivo pattern of binding of RAG1 and RAG2 to antigen receptor loci. Cell 141:419–431

Johnson K, Pflugh DL, Yu D, Hesslein DG, Lin KI, Bothwell AL, Thomas-Tikhonenko A, Schatz DG, Calame K (2004) B cell-specific loss of histone 3 lysine 9 methylation in the V(H) locus depends on Pax5. Nat Immunol 5:853–861

Johnston CM, Wood AL, Bolland DJ, Corcoran AE (2006) Complete sequence assembly and characterization of the C57BL/6 mouse Ig heavy chain V region. J Immunol 176: 4221–4234

Kim J, Sif S, Jones B, Jackson A, Koipally J, Heller E, Winandy S, Viel A, Sawyer A, Ikeda T, Kingston R, Georgopoulos K (1999) Ikaros DNA-binding proteins direct formation of chromatin remodeling complexes in lymphocytes. Immunity 10:345–355

Kosak ST, Skok JA, Medina KL, Riblet R, Le Beau MM, Fisher AG, Singh H (2002) Subnuclear compartmentalization of immunoglobulin loci during lymphocyte development. Science 296:158–162

Lawler AM, Lin PS, Gearhart PJ (1987) Adult B-cell repertoire is biased toward two heavy-chain variable-region genes that rearrange frequently in fetal pre-B cells. Proc Natl Acad Sci U S A 84:2454–2458

Liu H, Schmidt-Supprian M, Shi Y, Hobeika E, Barteneva N, Jumaa H, Pelanda R, Reth M, Skok J, Rajewsky K, Shi Y (2007a) Yin Yang 1 is a critical regulator of B-cell development. Genes Dev 21:1179–1189

Liu Y, Subrahmanyam R, Chakraborty T, Sen R, Desiderio S (2007b) A plant homeodomain in RAG-2 that binds Hypermethylated lysine 4 of histone H3 is necessary for efficient antigen-receptor-gene rearrangement. Immunity 27:561–571

Madisen L, Groudine M (1994) Identification of a locus control region in the immunoglobulin heavy-chain locus that deregulates c-myc expression in plasmacytoma and Burkitt's lymphoma cells. Genes Dev 8:2212–2226

Maes J, Chappaz S, Cavelier P, O'Neill L, Turner B, Rougeon F, Goodhardt M (2006) Activation of V(D)J recombination at the IgH chain JH locus occurs within a 6-kilobase chromatin domain and is associated with nucleosomal remodeling. J Immunol 176:5409–5417

Malin S, McManus S, Cobaleda C, Novatchkova M, Delogu A, Bouillet P, Strasser A, Busslinger M (2010) Role of STAT5 in controlling cell survival and immunoglobulin gene recombination during pro-B cell development. Nat Immunol 11:171–179

Malynn BA, Yancopoulos GD, Barth JE, Bona CA, Alt FW (1990) Biased expression of JH-proximal VH genes occurs in the newly generated repertoire of neonatal and adult mice. J Exp Med 171:843–859

Manis JP, Tian M, Alt FW (2002) Mechanism and control of class-switch recombination. Trends Immunol 23:31–39

Matthews AG, Kuo AJ, Ramon-Maiques S, Han S, Champagne KS, Ivanov D, Gallardo M, Carney D, Cheung P, Ciccone DN, Walter KL, Utz PJ, Shi Y, Kutateladze TG, Yang W, Gozani O, Oettinger MA (2007) RAG2 PHD finger couples histone H3 lysine 4 trimethylation with V(D)J recombination. Nature 450:1106–1110

Michaelson JS, Giannini SL, Birshtein BK (1995) Identification of 3' alpha-hs4, a novel Ig heavy chain enhancer element regulated at multiple stages of B cell differentiation. Nucleic Acids Res 23:975–981

Miller JP, Izon D, DeMuth W, Gerstein R, Bhandoola A, Allman D (2002) The earliest step in B lineage differentiation from common lymphoid progenitors is critically dependent upon interleukin 7. J Exp Med 196:705–711

Muller J, Kassis JA (2006) Polycomb response elements and targeting of Polycomb group proteins in Drosophila. Curr Opin Genet Dev 16:476–484

Nitschke L, Kestler J, Tallone T, Pelkonen S, Pelkonen J (2001) Deletion of the DQ52 element within the Ig heavy chain locus leads to a selective reduction in VDJ recombination and altered D gene usage. J Immunol 166:2540–2552

Nutt SL, Urbanek P, Rolink A, Busslinger M (1997) Essential functions of Pax5 (BSAP) in pro-B cell development: difference between fetal and adult B lymphopoiesis and reduced V-to-DJ recombination at the IgH locus. Genes Dev 11:476–491

Osipovich O, Milley R, Meade A, Tachibana M, Shinkai Y, Krangel MS, Oltz EM (2004) Targeted inhibition of V(D)J recombination by a histone methyltransferase. Nat Immunol 5:309–316

Pan PY, Lieber MR, Teale JM (1997) The role of recombination signal sequences in the preferential joining by deletion in DH–JH recombination and in the ordered rearrangement of the IgH locus. Int Immunol 9:515–522

Perlmutter RM, Kearney JF, Chang SP, Hood LE (1985) Developmentally controlled expression of immunoglobulin VH genes. Science 227:1597–1601

Ramon-Maiques S, Kuo AJ, Carney D, Matthews AG, Oettinger MA, Gozani O, Yang W (2007) The plant homeodomain finger of RAG2 recognizes histone H3 methylated at both lysine-4 and arginine-2. Proc Natl Acad Sci U S A 104:18993–18998

Reynaud D, Demarco IA, Reddy KL, Schjerven H, Bertolino E, Chen Z, Smale ST, Winandy S, Singh H (2008) Regulation of B cell fate commitment and immunoglobulin heavy-chain gene rearrangements by Ikaros. Nat Immunol 9:927–936

Roldan E, Fuxa M, Chong W, Martinez D, Novatchkova M, Busslinger M, Skok JA (2005) Locus 'decontraction' and centromeric recruitment contribute to allelic exclusion of the immuno-globulin heavy-chain gene. Nat Immunol 6:31–41

Saleque S, Singh M, Little RD, Giannini SL, Michaelson JS, Birshtein BK (1997) Dyad symmetry within the mouse 3' IgH regulatory region includes two virtually identical enhancers (C alpha3'E and hs3). J Immunol 158:4780–4787

Sayegh CE, Jhunjhunwala S, Riblet R, Murre C (2005) Visualization of looping involving the immunoglobulin heavy-chain locus in developing B cells. Genes Dev 19:322–327

Schwartz YB, Pirrotta V (2007) Polycomb silencing mechanisms and the management of genomic programmes. Nat Rev Genet 8:9–22

Sexton T, Bantignies F, Cavalli G (2009) Genomic interactions: chromatin loops and gene meeting points in transcriptional regulation. Semin Cell Dev Biol 20:849–855

Sollbach AE, Wu GE (1995) Inversions produced during V(D)J rearrangement at IgH, the immunoglobulin heavy-chain locus. Mol Cell Biol 15:671–681

Sridharan R, Smale ST (2007) Predominant interaction of both Ikaros and Helios with the NuRD complex in immature thymocytes. J Biol Chem 282:30227–30238

Stanton ML, Brodeur PH (2005) Stat5 mediates the IL-7-induced accessibility of a representative D-Distal VH gene. J Immunol 174:3164–3168

Su IH, Basavaraj A, Krutchinsky AN, Hobert O, Ullrich A, Chait BT, Tarakhovsky A (2003) Ezh2 controls B cell development through histone H3 methylation and Igh rearrangement. Nat Immunol 4:124–131

Subrahmanyam R, Sen R (2010) RAGs' eye view of the immunoglobulin heavy chain gene locus. Semin Immunol 22:337–345

ten Boekel E, Melchers F, Rolink AG (1997) Changes in the V(H) gene repertoire of developing precursor B lymphocytes in mouse bone marrow mediated by the pre-B cell receptor. Immunity 7:357–368

Thomas LR, Cobb RM, Oltz EM (2009) Dynamic regulation of antigen receptor gene assembly. Adv Exp Med Biol 650:103–115

Tsukada S, Sugiyama H, Oka Y, Kishimoto S (1990) Estimation of D segment usage in initial D to JH joinings in a murine immature B cell line. Preferential usage of DFL16.1, the most 5' D segment and DQ52, the most JH-proximal D segment. J Immunol 144:4053–4059

Urbanek P, Wang ZQ, Fetka I, Wagner EF, Busslinger M (1994) Complete block of early B cell differentiation and altered patterning of the posterior midbrain in mice lacking Pax5/BSAP. Cell 79:901–912

Vassetzky Y, Gavrilov A, Eivazova E, Priozhkova I, Lipinski M, Razin S (2009) Chromosome conformation capture (from 3C to 5C) and its ChIP-based modification. Methods Mol Biol 567:171–188

Vettermann C, Schlissel MS (2010) Allelic exclusion of immunoglobulin genes: models and mechanisms. Immunol Rev 237:22–42

Vosshenrich CA, Cumano A, Muller W, Di Santo JP, Vieira P (2003) Thymic stromal-derived lymphopoietin distinguishes fetal from adult B cell development. Nat Immunol 4:773–779

Vosshenrich CA, Cumano A, Muller W, Di Santo JP, Vieira P (2004) Pre-B cell receptor expression is necessary for thymic stromal lymphopoietin responsiveness in the bone marrow but not in the liver environment. Proc Natl Acad Sci U S A 101:11070–11075

Williams GS, Martinez A, Montalbano A, Tang A, Mauhar A, Ogwaro KM, Merz D, Chevillard C, Riblet R, Feeney AJ (2001) Unequal VH gene rearrangement frequency within the large VH7183 gene family is not due to recombination signal sequence variation, and mapping of the genes shows a bias of rearrangement based on chromosomal location. J Immunol 167:257–263

Wuerffel R, Wang L, Grigera F, Manis J, Selsing E, Perlot T, Alt FW, Cogne M, Pinaud E, Kenter AL (2007) S-S synapsis during class switch recombination is promoted by distantly located transcriptional elements and activation-induced deaminase. Immunity 27:711–722

Xu CR, Schaffer L, Head SR, Feeney AJ (2008) Reciprocal patterns of methylation of H3K36 and H3K27 on proximal vs. distal IgVH genes are modulated by IL-7 and Pax5. Proc Natl Acad Sci U S A 105:8685–8690

Yancopoulos GD, Alt FW (1985) Developmentally controlled and tissue-specific expression of unrearranged VH gene segments. Cell 40:271–281

Yancopoulos GD, Desiderio SV, Paskind M, Kearney JF, Baltimore D, Alt FW (1984) Preferential utilization of the most JH-proximal VH gene segments in pre-B-cell lines. Nature 311:727–733

Ye J (2004) The immunoglobulin IGHD gene locus in C57BL/6 mice. Immunogenetics 56:399–404

Zhang Z, Espinoza CR, Yu Z, Stephan R, He T, Williams GS, Burrows PD, Hagman J, Feeney AJ, Cooper MD (2006) Transcription factor Pax5 (BSAP) transactivates the RAG-mediated V(H)-to-DJ(H) rearrangement of immunoglobulin genes. Nat Immunol 7:616–624

Zhao Z, Tavoosidana G, Sjolinder M, Gondor A, Mariano P, Wang S, Kanduri C, Lezcano M, Sandhu KS, Singh U, Pant V, Tiwari V, Kurukuti S, Ohlsson R (2006) Circular chromosome conformation capture (4C) uncovers extensive networks of epigenetically regulated intra- and interchromosomal interactions. Nat Genet 38:1341–1347

Local and Global Epigenetic Regulation of V(D)J Recombination

Louise S. Matheson and Anne E. Corcoran

Abstract Despite using the same Rag recombinase machinery expressed in both lymphocyte lineages, V(D)J recombination of immunoglobulins only occurs in B cells and T cell receptor recombination is confined to T cells. This vital segregation of recombination targets is governed by the coordinated efforts of several epigenetic mechanisms that control both the general chromatin accessibility of these loci to the Rag recombinase, and the movement and synapsis of distal gene segments in these enormous multigene AgR loci, in a lineage and developmental stage-specific manner. These mechanisms operate both locally at individual gene segments and AgR domains, and globally over large distances in the nucleus. Here we will discuss the roles of several epigenetic components that regulate V(D)J recombination of the immunoglobulin heavy chain locus in B cells, both in the context of the locus itself, and of its 3D nuclear organization, focusing in particular on non-coding RNA transcription. We will also speculate about how several newly described epigenetic mechanisms might impact on AgR regulation.

Contents

1	Introduction	66
2	Epigenetic Mechanisms at the Level of the Igh Locus	68
	2.1 Histone Modifications	68
3	Non-Coding RNA Transcription	69
4	Intergenic Transcription	71
5	Antisense and Intergenic Transcription in the Igh Locus V Region	71

L. S. Matheson · A. E. Corcoran (✉)
Laboratory of Chromatin and Gene Expression, The Babraham Institute,
Babraham Research Campus, Cambridge, CB22 3AT, UK
e-mail: anne.corcoran@bbsrc.ac.uk

Current Topics in Microbiology and Immunology (2012) 356: 65–89
DOI: 10.1007/82_2011_137
© Springer-Verlag Berlin Heidelberg 2011
Published Online: 22 June 2011

6	Antisense and Intergenic Transcription in the Igh D Region	72
7	In vivo Function for Intergenic Transcription in V(D)J Recombination	73
8	Rag Binding	74
9	Do ncRNA Transcripts Play a Role in V(D)J Recombination?	75
10	Regulatory Elements Governing the Igh Locus	77
11	Nuclear Organization of V(D)J Recombination	78
12	The Role of Transcription in Nuclear Organization of V(D)J Recombination	80
13	Concluding Summary	83
References		84

1 Introduction

The role of the adaptive immune system is to produce B and T lymphocytes expressing an enormous repertoire of monoclonal high affinity antigen receptors (AgR) capable of responding to a huge diversity of foreign antigens. The first step in the generation of corresponding AgR diversity is V(D)J recombination of AgR loci. The AgR loci contain hundreds of genes in three groups, V (variable), D (diversity), and J (joining) that are assembled together in multiple combinations to generate a vast repertoire of sequences encoding unique antigen recognition motifs. The Rag1 and Rag2 recombinase enzymes, which catalyse the cleavage component of VDJ recombination, are expressed only in lymphocytes, thereby restricting V(D)J recombination to B and T cells. Recombination of individual AgR is further restricted to either B cells or T cells and specific developmental stages therein, by a dynamic and exquisitely co-ordinated set of epigenetic mechanisms. These processes collectively maintain AgRs in a closed chromatin conformation to ensure that Rag-mediated DNA double-strand breaks only occur in the appropriate AgR locus, open up appropriate AgR loci, and silence AgR loci again to prevent further recombination. For example, the immunoglobulin heavy chain locus (Igh) is never fully recombined in T cells. The Greek prefix 'epi' means 'over' or 'above'. Thus 'epigenetic' describes any change, over and above alteration in the DNA sequence itself, that influences the development of an organism (Holliday 1990). These mechanisms include DNA methylation, post-translational histone modifications, non-coding RNA (ncRNA) transcription, ATP-dependent chromatin remodeling, nuclear localisation, DNA looping, locus pairing. On the basis of one of the first discoveries of ncRNA transcripts in the genome, and before some of the other epigenetic mechanisms had been discovered, the much-cited Accessibility Hypothesis proposed some 25 years ago that ncRNA transcription in the Igh locus reflected an open chromatin state, and that differential chromatin accessibility of antigen receptor gene segments to the recombinase machinery regulated the lineage- and stage-specificity of recombination (Yancopoulos and Alt 1985; Stanhope-Baker et al. 1996). Thus this was one of the first epigenetic 'models'. Since then, we have come to understand that the antigen receptor loci, in particular Igh, Igk, TCRb, and TCR a/d are the largest loci

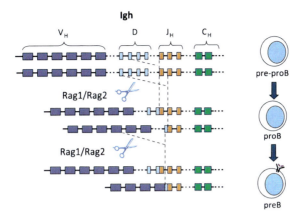

Fig. 1 Schematic diagram of V(D)J recombination of the mouse immunoglobulin heavy chain locus during B cell development. *Purple rectangles*: V (variable) genes; *grey rectangles*: D (diversity) genes; *orange rectangles*: J (joining) genes; *green rectangles*: C (constant) genes, not to scale. *Top line*: two homologous unrecombined Igh loci; *middle line*: both Igh loci D to J recombined using different D and J genes; *bottom line*: one of the two Igh loci has undergone further V to DJ recombination. On the *right*, the B cell stages at which these Igh locus configurations predominate are displayed. The pre-B cell has the preBCR on the cell surface

in the genome, and arguably the most complex, since before ultimate transcription of their successfully recombined VDJ gene, they must undergo at least one and often two rounds of V(D)J recombination, select one allele for expression and silence the other allele. They share this feature of mono-allelic expression with the imprinted genes and the olfactory receptor genes. V(D)J recombination itself requires simultaneous activation of 10s or 100s of genes, and targeted movement of those genes towards each other in the nucleus. This necessitates coordinated large-scale chromatin opening and physical DNA movement unprecedented in conventional gene expression.

In this review we will concentrate on the Igh locus as a mechanistic paradigm of epigenetic regulation of V(D)J recombination. The Igh locus consists of 195 V genes, 10 D genes, 4 J genes, and 8 constant region genes within a 3 MB sequence (Fig. 1). (Johnston et al. 2006) The Igh becomes accessible for recombination in pro-B cells, but here too epigenetic mechanisms dictate a strict order of accessibility and recombination: a D gene segment always recombines with a J segment before a V gene segment is appended to the DJ recombined gene segment—hence called D to J before V to DJ recombination (Fig. 1). Furthermore, D to J recombination occurs on both alleles but V to DJ recombination initially occurs only on one allele. If this recombination event produces an in-frame VDJ transcript that forms a functional Igh polypeptide, it is expressed on the cell surface in the context of the pre-BCR. It then signals back to the nucleus to prevent further V to DJ recombination on the second Igh allele, using epigenetic silencing mechanisms that largely reverse the opening mechanisms required for V to DJ recombination. This process is termed allelic exclusion, and ensures monoallelic expression of

AgR loci, which is vital to generate B cells with high affinity antibodies (Corcoran 2005). In this review, we will discuss features at the level of the Igh locus itself, including ncRNA transcription and regulatory elements, as well as the dynamic nuclear organization of these loci, while also highlighting important similarities and differences in the other AgR loci.

2 Epigenetic Mechanisms at the Level of the Igh Locus

DNA methylation was the first epigenetic process identified (Holliday 1990) and accordingly, the first epigenetic studies of the Igh locus investigated changes in DNA methylation, and concluded that DNA demethylation paralleled V(D)J recombination, and was necessary but not sufficient (Storb and Arp 1983; Engler and Storb 1999). However, lack of sequence information precluded detailed interrogation of V region CpG targets. Igkappa alleles undergo monoallelic demethylation, which appears to favor V–J recombination, but it is not clear whether this is a directive effect (Mostoslavsky et al. 1998). More recently it was shown that V(D)J recombination of Igh V genes can still occur when the DNA is methylated (Johnson et al. 2004), but further studies are required to establish whether and how this epigenetic mark plays a role in Igh recombination.

2.1 Histone Modifications

The picture is clearer for post-translational histone modifications that alter the association of DNA with nucleosomes. Dynamic changes in several histone modifications precede and follow recombination of individual Igh domains, and have been suggested to contribute to the order of chromatin opening and closing for V(D)J recombination. In pro-B cells preparing for Igh V(D)J recombination, alterations occur first over the DJ region, and include increases in DNase I sensitivity, histone acetylation, H3K4 methylation and nucleosome remodeling by BRG1 over DH and JH genes (Chowdhury and Sen 2001; Morshead et al. 2003; Maes et al. 2006). There appears to be an Eu-independent phase of intermediate chromatin opening, including removal of H3K9me2 marks, and acquisition of H3K4me2 modification. H3K9me2 removal is necessary for H3K9 acetylation, which recruits RNA PolII (Chakraborty et al. 2009). Subsequently, repressive histone H3K9 dimethylation is lost, and active histone modifications including histone H3 and H4 acetylation are acquired across the V regions in a step-wise manner. Following V(D)J recombination, active histone modifications are reduced (Chowdhury and Sen 2001, 2003; Maes et al. 2001; Johnson et al. 2003; Morshead et al. 2003; Johnson et al. 2004).

Notably, candidate gene approaches have identified, for example H3K4me2, at individual *V* genes or small gene families, because they have been specifically interrogating those genes (Johnson et al. 2004). In contrast, locus-wide analysis of H3K4me3, a signature of active promoters, has revealed surprisingly few peaks of enrichment in the V region, given that each V gene has a bona fide promoter, and indeed the peaks of enrichment appear to be in intergenic regions (Malin et al. 2010). However it is, perhaps uniquely, difficult to interpret this sort of data for AgR loci, since a fundamentally relevant question still has not been answered. In the Igh for example, does the V region of an individual Igh allele, containing 195 genes, have all, or just a few, or even only one of its V genes accessible for recombination at the appropriate time? If for example, the answer is only one, and this occurs in a stochastic fashion such that only one allele in 195 has this gene active, then in a mixed cell population, this would appear as the background in competitive genome-wide mapping studies. To address this issue, technically challenging approaches including single cell CHIP-seq may need to be employed.

Much of this data, while strongly supporting a key role for dynamic histone modifications in controlling accessibility for V(D)J recombination, is nevertheless correlative, and awaits loss of function studies to determine the hierarchical importance of individual histone modifications. The importance of one such repressive histone modification, H3K27 trimethylation, was functionally suggested by a targeted deletion of the H3K27 histone methyltransferase, ezh2, which caused reduced recombination of D-distal V genes, although it was unclear whether this was due to a direct loss of H3K27 trimethylation at the Igh locus itself, or an indirect effect due to activation of an unknown factor (Su et al. 2003). This conundrum was resolved by locus-wide mapping of H3K27me3, which showed that, somewhat counter-intuitively since 3' V genes recombine slightly earlier than 5' V genes, H3K27 trimethylation is normally confined to V genes at the 3' end of the V region, while H3K36 trimethylation, which characterizes transcribing gene bodies predominates at the 5' end of the V region (Xu et al. 2008). The ensuing model proposed that H3K27 trimethylation inactivates the 3' V genes, to enable the 5' V genes to recombine. Thus its loss may confer a sustained recombination advantage on the 3' V genes, with the perceived outcome as reduced 5' V gene recombination. Indeed, loss of ezh2 has also been reported to cause reduced DNA looping (discussed below), but this may be secondary to the preferential usage of 3' V genes, such that looping simply occurs less frequently because recombination of 3' V genes has occurred more frequently. Furthermore, the well-documented bias in favor of 3' V gene usage in V(D)J recombination in fetal liver, coincides with an absence of H3K27me3 marks at 3' V genes in these B cells (Xu et al. 2008).

3 Non-Coding RNA Transcription

In the burgeoning epigenetic field, non-coding RNA transcription has only recently gained recognition as an epigenetic mechanism. This is largely due to recent genome-wide transcriptional analyses showing that the vast majority of transcribed

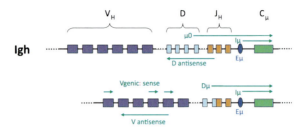

Fig. 2 Dynamic changes in non-coding RNA transcription of the Igh locus during V(D)J recombination. Transcripts are represented by *green arrows*; intronic enhancer Eu: *dark blue oval*. *Top line*: unrecombined Igh locus; *bottom line*: DJ recombined Igh locus poised for V to DJ recombination

mammalian genomic sequences are non-protein-coding (Carninci et al. 2005; Kapranov et al. 2007; Mercer et al. 2009). These findings have overturned the 'central dogma' that DNA makes RNA makes protein, and dramatically changed how we think about genome regulation. These non-coding RNAs fall into several distinct categories, all of which play previously unsuspected key roles in gene regulation (Mercer et al. 2009). Here we will focus on long non-coding RNAs, a term conventionally used to define any ncRNA transcript of more than 200 bp. In the mouse genome, long ncRNAs number approximately 30,000 (Carninci et al. 2005) and they constitute the majority of transcribed sequences in the human genome (Kapranov et al. 2007).

Historically, the Igh locus was the one of the first loci shown to express non-coding RNA transcripts (Yancopoulos and Alt 1985). Strikingly, these transcripts were specifically expressed from Igh domains poised for V(D)J recombination. This transcription was originally termed 'sterile' or 'germline' to distinguish it from coding transcription from V(D)J recombined genes, but we will refer to it as non-coding (nc) in compliance with the current terminology. The first ncRNA transcripts initiate on unrecombined Igh alleles, from the intronic enhancer Eμ (Iμ transcript) (Lennon and Perry 1985), and from the PDQ52 promoter/enhancer, upstream of the most J-proximal D gene segment, DQ52 (μ0 transcript) (Thompson et al. 1995) (Fig. 2). After D to J recombination, the *DJ* gene segment produces Dμ transcripts (Reth and Alt 1984) and ncRNA transcription initiates over the V genes (Fig. 2) (Yancopoulos and Alt 1985; Corcoran et al. 1998). Similar non-coding RNA transcripts have been identified in all antigen receptor loci poised for recombination (Corcoran 2005). There has been much debate regarding whether this genic ncRNA transcription has a bona fide function, as opposed to simply being a passive output of gene promoters activated for V(D)J recombination. Numerous in vitro studies have addressed the question, some supporting a directive role, while others have found no relevant function (Abarrategui and Krangel 2009), but to date the function of this genic transcription has not been interrogated in vivo.

4 Intergenic Transcription

Most studies of ncRNA transcription in the AgR loci have focused on recombining gene segments. In numerous multigene loci, including β-globin, T helper 2, and the MHC complex, an additional process, termed intergenic transcription, processes through silent chromatin to generate extensive domains of modified chromatin that encompass active genes and their regulatory elements, and facilitate additional specific chromatin unfolding over genes (Gribnau et al. 2000; Masternak et al. 2003; Bernstein et al. 2005). Indeed intergenic transcription through enhancers is essential for activation of numerous target genes (Ho et al. 2006; Kim et al. 2010). Chromatin reorganization is achieved in part by the recruitment of histone acetyltransferases (HATs), histone methyltransferases (HMTs), and SWI/SNF ATP-dependent chromatin remodeling complexes by RNA polymerase II (PolII) as it progresses through the chromatin (Wilson et al. 1996; Ng et al. 2003). Furthermore, transcription promotes replication-independent histone exchange for variant histone H3.3, enriched with active modifications (Mito et al. 2005). Intergenic transcription is thought to initiate in silent chromatin with the aid of 'pioneering factors', although these have not yet been found (Orphanides and Reinberg 2000). This transcription-dependent higher order chromatin remodeling has been proposed to allow large developmentally regulated loci, such as the AgR loci, to loop out of their chromosome territories and become actively transcribed (Volpi et al. 2000).

5 Antisense and Intergenic Transcription in the Igh Locus V Region

The chromatin remodeling processes initially discovered in the Igh V region discussed above, were confined to V genes. These small-scale alterations are separated by large intergenic regions (Johnston et al. 2006), and thus are insufficient to remodel silent chromatin throughout the locus in non-B cells (Johnson et al. 2004). We hypothesized that a large-scale process such as intergenic transcription would be necessary to make all of the V genes accessible efficiently. We analyzed transcription throughout the Igh VH region, using RNA-FISH to visualize primary transcripts on individual alleles in single cells. These studies revealed that intergenic transcription occurs throughout the Igh V region in B cell progenitors. It is absent on germline alleles that have not yet undergone DH to JH recombination, is expressed on the majority of DHJH recombined alleles, and is lost following V to DJ recombination (Bolland et al. 2004) (Fig. 2). Thus it is strictly developmentally regulated. Furthermore, these transcripts occur at relatively high frequency, similar to housekeeping genes, which is very unusual for a non-coding transcript, and RNA-FISH signal patterns demonstrate extensive transcription on individual alleles. Collectively, these data indicate that Igh V region intergenic transcription is a large-scale functional mechanism.

Surprisingly, this intergenic transcription occurred exclusively on the antisense strand with respect to the Igh *V* genes, which are all transcribed in the same orientation, towards the DJ region. Classically associated with imprinted loci, in which it silences gene expression in cis on one allele to facilitate monoallelic expression (Sleutels et al. 2002; Nagano et al. 2008), antisense transcription is now understood to play a myriad of roles, both activating and silencing in higher eukaryotes. A large proportion of transcription units have overlapping sense and antisense transcription, which is often coordinately regulated (Katayama et al. 2005). Antisense transcription is sometimes antagonistic at coding gene promoters (He et al. 2008), but on the other hand, it is also required to activate neighboring genes in multigene clusters, by disrupting interaction with repressive PcG complexes and recruiting activating Trithorax (TrX) complexes (Rinn et al. 2007; Sessa et al. 2007).

Since V antisense transcription occurs only after D to J recombination has occurred, and occurs on both DJ recombined alleles, at a stage when the V region must now open up for V to DJ recombination, this favors a role for antisense intergenic transcription in opening up the VH region on both DJ recombined alleles which are poised for V to DJ recombination. This study was the first report of intergenic transcription in V(D)J recombination. We proposed that this large-scale transcription remodels the entire Igh VH region, thus facilitating subsequent more focused changes in chromatin structure over *V* genes before V to DJ recombination (Bolland et al. 2004).

6 Antisense and Intergenic Transcription in the Igh D Region

These discoveries raised the question of whether similar transcription preceded other V(D)J recombination events. We subsequently found that antisense intergenic transcription also occurs throughout the DH (60 kb) and JH regions of the mouse Igh locus, specifically on germline alleles poised for DH-to-JH recombination, again at a stage when the chromatin must be made accessible. Notably, it initiates immediately upstream of, and is regulated by the intronic enhancer $E\mu$ (Bolland et al. 2007). Targeted deletion of $E\mu$ causes a defect in the DH-to-JH recombination (Perlot et al. 2005; Afshar et al. 2006), but it is not yet clear how this occurs mechanistically, since several processes are affected. Deletion of $E\mu$ results in loss of both $I\mu$ (Afshar et al. 2006) and D antisense transcription, over the 60 kb region (Bolland et al. 2007). This suggests that $E\mu$ controls DH-to-JH recombination at least in part by activating antisense transcription (Fig. 2). Subsequent acquisition of H3K4 trimethylation requires $E\mu$, and this modification is recruited by RNA *Pol*II, supporting a directive, hierarchical role for transcription, regulated by Eu (Chakraborty et al. 2009). In vitro studies showing that germline transcription of Ig loci precedes acquisition of post-translational histone modifications, support this model (Xu and Feeney 2009). Interestingly, SWI/SNF ATP-dependent chromatin remodeling complexes are required for V region antisense transcription, and $E\mu$ dependent $I\mu$, μo, and

D antisense transcription, suggesting that SWI/SNF remodeling may supply the 'pioneering' activity required to initiate intergenic transcription in the closed chromatin of Igh domains (Osipovich et al. 2009).

This transcription occurs over the 5′DFL16 D gene, as well as at the middle DSP D genes. Nevertheless, an alternative hypothesis has been proposed, that antisense transcription may contribute to repression of the middle DSP genes by formation of dsRNA and heterochromatinization, although no dsRNA was detected (Chakraborty et al. 2007), nor any sense transcription over DSP genes (Bolland et al. 2007). Strikingly, the only region where both sense and antisense transcripts have been observed is over the DQ52 gene and the J genes, which are coordinately up-regulated by Eμ (Bolland et al. 2007), and in a hyper accessible chromatin domain (Maes et al. 2006), implying that they do not produce dsRNAs that lead to heterochromatin. Clarification of these alternative models will require removal of this transcription in vivo to unambiguously determine its functional role.

7 In vivo Function for Intergenic Transcription in V(D)J Recombination

Importantly, proof that intergenic transcription is functionally required for V(D)J recombination has been provided by gene targeting studies that ablated intergenic transcription in the Tcra locus J region in vivo (Abarrategui and Krangel 2006; 2007). Two different effects were observed. Loss of transcription over genes lacking their own adjacent promoters inhibited their recombination, suggesting that upstream intergenic transcription reading through these gene segments was required to make them accessible, and indeed loss of this transcription prevented Rag1 binding to these J genes (Ji et al. 2010a). Moreover, H3K4me3 marks were also lost, which may reduce binding of RAG2 recombinase (Matthews et al. 2007). Conversely, downstream J genes with adjacent promoters were activated, suggesting that intergenic transcription normally exerted transcriptional interference over these promoters, which may contribute to ordering of recombination.

Notably, intergenic transcription in the Tcra locus originates from the sense strand. Since strand origin of transcription is unlikely to be important for general chromatin opening, these studies support the model that antisense transcription contributes to chromatin opening in the Igh locus. However, there are key differences that preclude a direct comparison. In the TCRa locus, intergenic transcription antagonizes genes with promoters, while in the Igh V region, every V gene has its own transcribing promoter (Johnston et al. 2006), but arguably if antisense transcription inhibits these, it would be by a different mechanism. Secondly, the TCRa J array has 61 J genes within 65 kb. In contrast, the 2.5 MB Igh V region, with average distances of 10–20 kb between V genes requires chromatin unfolding on a much larger scale, which importantly involves DNA looping. The unidirectionality of antisense transcription may favor this process. We are currently testing this model by transcription ablation in vivo (Andy Wood et al. manuscript in preparation).

Importantly, an in vivo 'gain of function' model has recently been generated in the Igh locus that directly supports a functional role of antisense transcription in Igh V(D)J recombination. The 100 kb sequence between the Igh V and D regions was deleted, to functionally test the hypothesis that it contained regulatory elements that control ordered VDJ recombination (Giallourakis et al. 2010) (see below). The consequences included de novo appearance of antisense transcripts from D-proximal V genes in T cells, which directly correlated with de novo V–D recombination events using those V genes in T cells, which normally do not undergo V to DJ recombination. Furthermore, T cells normally undergo limited Igh D to J recombination, but the deletion caused a dose-dependent increase in D antisense transcription, coupled with increased frequency of D to J recombination events. These findings validate our model that antisense transcription has an directive activating influence on V(D)J recombination.

8 Rag Binding

All of the above mechanisms have a common goal—to make the AgR accessible for Rag binding. Recent studies have clarified the nature and extent of Rag binding, both at AgR loci, and throughout the genome. The JH region and the DQ52 gene form a separate even more open chromatin domain to the rest of the DH region (Chowdhury and Sen 2001; Morshead et al. 2003; Maes et al. 2006), and accordingly the DQ52 gene is preferentially used in early DH-to-JH recombination. Recent genome-wide CHIP-seq analysis of Rag1 and Rag2 binding has shown remarkably focused binding of both recombinases at J genes and J-proximal D genes (e.g., DQ52) in several antigen receptor loci, compared with surprisingly low binding frequency at V genes, suggesting that the highly open chromatin structure may promote preferential Rag targeting to J genes and contribute to recombination order (Ji et al. 2010b). However, the caveats previously mentioned for quantification of histone modifications in bulk populations also apply to a simple interpretation that the Rags bind only to the J genes. Intriguingly, Rag2 was required for Rag1 binding to the Igh J region, but not to the other AgRs, although the reason for this difference is unknown. Rag2, which binds H3K4me3 marks (Matthews et al. 2007), was also shown to bind to over 99% of H3K4 trimethylated sites in the genome, suggesting that its binding is not particularly specific to the J regions. Genome-wide binding of Rag1 generally did not overlap with Rag2 binding, and in this case, an open chromatin structure and the presence of an RSS were the critical components.

These recent studies prompt a reconsideration of the Accessibility Hypothesis, which originally proposed that keeping a particular AgR silent prevents Rag access for recombination. Indeed this remains true for the TCRs in B cells and vice versa, but inherent in this was the notion that the Rags only bound AgRs. Extraordinarily, since Rag2 binds promiscuously throughout the genome, the key question for the future, both to understand V(D)J recombination, and genomic integrity is: What provides the specificity that prevents Rag-mediated DNA DSBs occurring throughout the genome? The spotlight has now fallen on Rag1, both as the chief orchestrator of binding specificity of the Rag complex to the AgR loci, and also as the gate-keeper of genomic integrity, since fortuitously both Rags are required to induce aberrant recombination events. Thus it is now more important than ever to answer the long-standing question: how does the Rag complex bind? Rag1 first, Rag2 first, or a preformed complex of the two?

The accessibility hypothesis has only recently been tested directly due to the recent generation of antibodies against Rag1 and Rag2, previously notoriously difficult to generate. These studies directly analyzed Rag1 binding to the TCRa and TCRb loci in vivo in mouse models of promoter, enhancer, and transcription dysfunction, and demonstrated that enhancers were required for global recruitment of Rag1, and that promoters and intergenic transcription were required for local recruitment to individual genes (Ji et al. 2010a). Overall, these studies validate the Accessibility Hypothesis by directly showing for the first time the effect of loss of these aspects of chromatin organization on Rag1 binding.

The original premise of the Accessibility Hypothesis was that locus opening and Rag binding were two separate, sequential events. Now it appears they are directly and inextricably linked, since one of the histone modifications that opens up the loci has the dual function of recruiting Rag2. Interestingly, Rag1 has its own histone modifying activity. Its N-terminal RING domain interacts with and promotes monoubiquitylation of histone H3, which appears to be required for the joining step in vivo, making an important link between the cutting and joining steps of this reaction (Grazini et al. 2010). Together these observations pose the question: Is there a unique histone code for V(D)J recombination?

9 Do ncRNA Transcripts Play a Role in V(D)J Recombination?

If transcription is important for V(D)J recombination, several questions follow— How does it act? Is transcription 'one-off' to initate chromatin opening, or continuous to maintain open chromatin? To initiate unfolding of silent chromatin, intergenic transcription might only need to process through a DNA domain once or twice. However, continuous intergenic transcription is often required to maintain open chromatin by actively preventing binding of repressive PRC H3K27 HMT and by recruitment of activating TrX H3K4 HMT complexes (Schmitt et al. 2005; Sessa et al. 2007; Rinn et al. 2007). We find that intergenic transcription is more frequent in the $5'$ half of the V region (Adam Bowen, manuscript in preparation). Conversely, H3K27me3 modification is confined to the $3'$ end of the V region (Xu et al. 2008). An intriguing possibility is that continuous intergenic transcription in the $5'$ half prevents this part of the V region from acquiring the repressive H3K27me3 mark, and thus keeps the chromatin in an open state permissive for DNA looping (discussed below).

In addition to the process of large-scale intergenic transcription, numerous studies have revealed roles for intergenic RNAs, which point to multiple roles in gene regulation, including activation of accessory proteins, transport of transcription factors (Ponting et al. 2009), and recruitment of repressive chromatin modifiers in cis (Nagano et al. 2008; Zhao et al. 2008). Recent genome-wide analyses have uncovered large classes of novel, conserved intergenic transcripts. These include the long intervening non-coding RNAs or LincRNAs. These transcripts have a highly conserved histone modification signature, consisting of a peak of H3K4me3, with an adjacent stretch of H3K36me3, that occurs 1000s of times within intergenic regions at least 5 kb from annotated genes in both mouse and human genomes (Guttman et al. 2009). These marks are normally associated with coding gene promoters and gene bodies respectively. Characterization of lincRNAs has revealed multiple regulatory functions (Guttman et al. 2009). One of the best understood is the 2.2 kb (HOTAIR HOX Antisense Intergenic RNA). HOTAIR transcription upstream of and antisense to its adjacent *HoxC* genes maintains a domain of open chromatin that stops spreading of heterochromatin into the region, at least in part by recruiting H3K4me3 modifications (Rinn et al. 2007). Conversely, the HOTAIR transcript itself appears to repress transcription from the HoxD cluster in trans on a separate chromosome, by acting as a 'molecular scaffold' to recruit and organize the concerted actions of two repressive chromatin-modifying enzymes: the H3K4 histone demethylase, LSD1 (Tsai et al. 2010), and PRC2 (Rinn et al. 2007). Indeed 20% of lincRNAs recruit PRC2, (Khalil et al. 2009) and large non-coding RNAs may represent a trafficking system that guides ubiquitous chromatin modifying complexes to specific targets, both in cis and in trans (Koziol and Rinn 2010). Such lincRNAs can have many targets (Huarte et al. 2010). It will be interesting to determine whether some of the other 80% of lincRNAs transport activating chromatin modifying complexes. The RNA does not bind to homologous sequence but the RNA secondary structure may play a key role in binding of complexes and recognition of DNA sequences (Koziol and Rinn 2010). Transcription from a single non-coding locus can thus have both activating and repressive roles in gene expression, both in cis and in trans.

A function for ncRNA transcripts has not yet been demonstrated in any AgR locus. Since antisense intergenic transcripts are more abundant in the 5' half of the V region, and H3K27me3 is recruited exclusively to the 3' end, (Xu et al. 2008) an alternative model to the one above is that transcripts from the 5' end may guide the PRC2 complex to the 3' end in cis, in a similar manner to Airn (Nagano et al. 2008). Alternatively, ncRNAs could act in trans to downregulate the homologous Igh allele in a similar manner to HOTAIR from HoxC silencing HoxD (Rinn et al. 2007), since in both cases there is an imperative to silence another allele. If ncRNA transcripts have an activating function, they might recruit activating histone modifiers for further chromatin opening at specific sites. To answer the question of whether transcription or the transcripts are important, in vivo ablation of transcription will not be enough. This will require nuclear knockdown of the ncRNA transcripts, a technique that has had variable success, and will be challenging for the very long ncRNAs in the Igh.

10 Regulatory Elements Governing the Igh Locus

Apart from Eμ, which as discussed above is required for efficient D to J recombination (Perlot et al. 2005; Afshar et al. 2006), there is a surprising paucity of regulatory elements identified to date in the Igh locus. Two candidate enhancers, at the PDQ52 gene and 5' of the V region failed to demonstrate a major role in V(D)J recombination in vivo (Nitschke et al. 2001; Pawlitzky et al. 2006; Perlot et al. 2010). Thus the search for a V region regulatory element continues unabated.

Intriguingly, a large family of conserved elements, termed Pax5-activated intergenic repeat (PAIR) elements have recently been identified. Eleven of these PAIRs are located upstream of V_H3609 genes in the distal part of the Igh V region. They bind Pax5, E2A, CTCF, and Rad21, all of which are implicated in DNA looping, and exhibit active histone modifications and associated antisense transcription, which are dependent on Pax5 and confined to pro-B cells (Ebert et al. 2011). Representative members display promoter or enhancer activity in vitro and the authors propose that these elements may be the elusive elements proposed to direct long-range interactions of the distal V region with the DJ region to facilitate V to DJ recombination, perhaps by forming the base of rosette loops (discussed below) (Jhunjhunwala et al. 2008). Functional analysis of these elements will be an important avenue for future studies.

Importantly, in addition to individual reports that transcription through enhancers is essential for activation of target genes (Ho et al. 2006), genome-wide studies of conserved RNAs sharing features with lincRNAs, have demonstrated that a large subset have enhancer-like function in human cells. They can operate up and downstream of gene targets from considerable distances and on the opposite strand (Orom et al. 2010). Importantly, the functionality of the RNAs themselves, rather than the process of their transcription, was established by ncRNA knockdown. Moreover, genome-wide studies of neuronal enhancers have revealed widespread enhancer-associated transcripts or eRNAs that directly correlated with expression of target genes (Kim et al. 2010). It is tempting to speculate that some of the antisense transcripts in the Igh V region may play the role of the elusive enhancer.

In addition to enhancers of V(D)J recombination, elements that contribute to ordered recombination in other ways have been predicted to be present in AgR loci. Several recent studies have focused on insulators, which possess both barrier (boundary) function to prevent spreading of histone modifications (e.g., those associated with heterochromatin), and enhancer-blocking activity that protects promoters from the activity of enhancers or silencers. This almost invariably requires CTCF binding, which has been proposed to isolate chromatin domains by facilitating looping out of DNA (Phillips and Corces 2009). Insertion of a V gene immediately upstream of the mouse Igh D region in vivo disrupted ordered recombination, suggesting that it had circumvented an unknown upstream insulator (Bates et al. 2007). We recently characterized the 96 kb V–D intergenic sequence, strategically located between the mouse Igh V and D regions, and found that it contained several conserved lineage-specific DNase hypersensitive sites, including two adjacent sites, HS4 and HS5, close to the D region. These elements bound CTCF, and had potent enhancer-blocking activity in vivo. The two CTCF sites were also previously identified in a CHIP-chip microarray (Degner et al. 2009). Furthermore, HS4 and HS5 are situated at an interface between active and repressive histone modifications, indicating boundary function. This location also coincides with a dramatic decrease in the antisense transcription which has progressed several kb upstream of the first D gene (Featherstone et al. 2010). We have proposed that these insulators prevent D antisense transcription from progressing further towards the V region and remodeling V region chromatin. These findings lend mechanistic support to a recent study that combined DNA-FISH with mathematical modeling to provide the first model of dynamic structural alterations in the Igh locus at sequential stages of B cell development (Jhunjhunwala et al. 2008). The authors showed that a DNA sequence adjacent to the HS4/HS5 insulators, as well as Eu and the intervening DJ region, are sequestered adjacent to the 3'RR by DNA looping in uncommitted pre-pro-B cells that have not undergone D to J recombination. They suggest that this geographical separation prevents the participation of V genes in the first round of recombination. We propose that HS4 and/or HS5, mediate this separation by DNA looping to interact with CTCF sites in the 3'RR (Garrett et al. 2005). In $Rag^{-/-}$ pro-B cells poised for V(D)J recombination, the Igh locus conformation alters to re-position the DJ region and its upstream V–D sequence proximal to the V region (Jhunjhunwala et al. 2008). HS4 and HS5 may facilitate this repositioning towards the V region in pro-B cells, perhaps by engaging with CTCF binding sites in the V region (Degner et al. 2009) (discussed below).

The V–D knockout described above (Giallourakis et al. 2010) provides important in vivo validation of our model, since it results in loss of the two insulators in vivo, leading to de novo antisense transcription and recombination in the 3' end of the Igh V region in T cells, strongly suggesting that these insulators normally prevent transcriptional read-through and activation of 3'V genes in lymphocytes that have not yet undergone D to J recombination. Loss of the V–D region in T cells also increased antisense transcription in the D region, and D to J recombination. This suggests that the insulators or other elements in the V–D region also modulate the DJ region in T cells, only a minority of which undergo Igh D to J recombination. V and D antisense transcription were also increased in $Rag2^{-/-}$ pro-B cells, and it will be important to determine whether this reflects a modulating effect of the V–D region on Igh V to DJ or D to J recombination in their normal B cell context.

11 Nuclear Organization of V(D)J Recombination

Both enhancers and insulators engage in looping and 3D DNA movement within the nucleus. Recent advances in the generation of genome-wide data sets and methods that interrogate the 3D association of gene loci with each other have transformed our understanding of gene regulation, and revealed layers of complexity and sophistication never before envisaged. The nucleus, often considered to be an unorganized heterogeneous mixture of DNA and associated proteins, is proving to be a highly organized cellular organelle containing numerous specialized sub-compartments (Spector 2003). DNA is generally constrained within chromosome territories, which maximizes DNA packaging efficiency (Cremer and Cremer 2010). However, chromatin retains the flexibility to easily fold and unfold individual genomic loci, and to spatially segregate open and closed chromatin into distinct compartments (Lieberman-Aiden et al. 2009). Furthermore, genomic loci undergo numerous chromosomal associations, both in cis and in trans (Schoenfelder et al. 2010b). Thus, long-range chromosomal associations between genomic regions, and their repositioning in the 3D space of the nucleus, are now considered to be key contributors to the regulation of gene expression (Schoenfelder et al. 2010a). Thus one of the simple but profound challenges facing biologists studying gene regulation today is to forget the imprinted picture of DNA as a long, straight, isolated piece of string with genes dotted along it, and squiggly RNA transcripts passively emanating from these dots, but rather to think of it as a highly folded and dynamically changing convoluted structure, intertwined with many other long pieces of strings, from which RNAs emerge with many diverse functions. Nowhere is this more imperative than in the Ig loci, where complex 3D looping is the modus operandi!

The AgR are a classic paradigm for DNA movement, since V(D)J recombination requires juxtaposing of distal gene segments. In non-B cells the Igh is sequestered at the inner nuclear membrane, tethered by the 5 end of the V region (Yang et al. 2005), which may limits its accessibility for transcription and recombination (Reddy et al. 2008). In pro-B cells the locus relocates to the nuclear interior, which is enriched in permissive euchromatin (Kosak et al. 2002). The DJ region repositions first, and this order is very likely to contribute to ordered recombination. D to J recombination and V to DJ recombination of DJ-proximal V genes region proceed without further large-scale DNA movement. However, since the middle and distal V genes are spatially distant from the DJ region, they require an additional 3D conformational change to bring them close enough to the DJ region for a V gene to synapse with a DJ gene segment for recombination. This is achieved by DNA looping, or contraction, which brings distal V genes in close proximity to the DJ region (Roldan et al. 2005; Sayegh et al. 2005), such that both proximal and distal V regions are at a similar distance from the DJ region in 3D nuclear space and thus equally available for recombination (Jhunjhunwala et al. 2008). Several models have been proposed to depict the kinds of loops formed. The most sophisticated, which combines extensive DNA-FISH measurements with mathematical modeling and trilateration, predicts that the Igh V region chromatin fibre folds into bundles of loops, in a rosette-like conformation (Jhunjhunwala et al. 2009), and that these loops form around a hollow containing the DJ region (Lucas et al. 2011). Several transcription factors are required for looping, although it remains unclear whether they play direct roles at the locus, or act indirectly. These include Pax5, YY1, ikaros (Fuxa et al. 2004; Liu et al. 2007; Reynaud et al. 2008).

Furthermore, based on the recent findings of CTCF/cohesin binding sites in strategic locations throughout the Igh V region (Degner et al. 2009) and our own work (Ciccone et al. unpublished results), and within the V–D intergenic region, there is also a strong possibility that CTCF and cohesin participate in looping. The CTCF sites found invariably close to the 3 RSS at Igh 3'V genes may facilitate looping of these 3 genes to provide 'equal opportunity for all' (Lucas et al. 2011). The CTCF sites in the distal part of the V region are mostly intergenic and their potential mode of action is less clear. However, HS4 and HS5 in the V–D region (Featherstone et al. 2010) are strategically placed to provide a focal point for looping of these distal CTCF sites adjacent to the DJ recombined gene segment. Notably, DNA looping still occurs in the absence of Rags, although at reduced frequency, suggesting that large-scale movement of the V regions in cis to facilitate V to DJ recombination is not dependent on the Rag complex. This is important when considering the long-standing question: How do the Rags bring the RSSs from two gene segments together? The 'capture model' (Jones and Gellert 2002) proposes that Rags binds one RSS, then 'captures' the second. If the Rags are not required for large-scale looping, this suggests that such capture takes place at a local level after the V region has looped to the DJ region, rather than playing an active part in large-scale Igh locus movement.

An additional layer of nuclear organizational control has recently been discovered, termed locus pairing (Hewitt et al. 2009). Previously it was assumed that V to DJ recombination occurred independently on the two Igh alleles in separate regions of the nucleus and that the asynchronous choice of one allele that eventually produced a productive Igh rearrangement was the stochastic result of an inherently inefficient process. However, this study demonstrates that the homologous Igh alleles often reposition adjacent to each other, and that this is dependent on the Rag complex, and sets the stage for V(D)J recombination. Although this level of inter-locus coordination, and indeed this additional movement within the nucleus is surprising, this kind of pairing is not without precedent, since the two X chromosomes pair briefly before X-inactivation (Bacher et al. 2006). In both cases there is a common goal of silencing one of the homologous alleles. V(D)J recombination can still occur without pairing and it will be important to determine how and why the two recombination contexts differ.

12 The Role of Transcription in Nuclear Organization of V(D)J Recombination

Where does non-coding RNA transcription fit in? Rather than taking place randomly in the nucleus, transcription is concentrated in metastable pre-formed subnuclear foci of active RNA *Pol*II complexes, termed transcription factories (Mitchell and Fraser 2008). Factories transcribe several genes simultaneously, and these genes can be up to 40 MB apart in cis, or on separate chromosomes (Osborne et al. 2004, 2007). Genes co-transcribed in the same factory often have related functions, which collectively constitute a functional transcriptional 'interactome' (Schoenfelder et al. 2010b). Importantly for V(D)J recombination, the Igh locus must move in 3D nuclear space to find a transcription factory in which non-coding RNA transcription can take place, and we propose that juxtaposing of distal transcribing parts of the Igh locus in a transcription factory directs them into a favorable setting for V(D)J recombination.

First, we have shown that the non-coding transcript Iμ, generated from the Igh Eμ intronic enhancer, is transcribed almost continuously in pro-B cells undergoing V(D)J recombination (Bolland et al. 2004), and throughout B cell development (Osborne et al. 2007). This frequency of transcription only occurs in a minority of coding genes, termed supergenes, which tend to be important markers of lineage-specific function (Fraser 2006). It is unprecedented in non-coding genes, and thus Im may be the first 'super-intergene' identified (Fraser 2006). Importantly, this means that it is constantly associated with a transcription factory (Fig. 3) (Osborne et al. 2004). In mature activated B cells, this continuous Iμ transcription results in frequent co-localisation of the Igh with the transcribing proto-oncogene myc in a transcription factory, and this proximity of actively transcribing and open chromatin may contribute to the Igh-myc translocations characteristic of Burkitt's lymphoma (Osborne et al. 2007). Furthermore, enhancers can relocate genes away from the nuclear periphery by recruiting them to a transcription factory (Ragoczy et al. 2006). Eμ may thus recruit the DJ region to a transcription factory, where it is retained by continuous Iμ transcription, providing a stable focal point for D to J recombination (Fig. 3). Second, thereafter DNA looping of V genes in close proximity for V to DJ recombination may occur at least in part because large parts of the non-coding V region dynamically engage with this transcription factory. Consequent co-association of a particular individual V gene with the DJ region in the same factory may position it favorably for V to DJ recombination (Fig. 3). Furthermore, 'specialized' transcription factories are formed by frequent association of common transcription factors that co-ordinate transcription of co-regulated genes (Schoenfelder et al. 2010b). The AgR loci are a classic example of co-regulated genes that might preferentially co-associate in a transcription factory. As mentioned above, CTCF may nucleate a 'recombination factory' of 3' V genes (Lucas et al. 2011). Since the 5' V genes have a common requirement for several B cell-specific factors that regulate histone modifications, ncRNA transcription (Stat5, ikaros, ezh2) (Bertolino et al. 2005; Reynaud et al. 2008) or looping (Pax5, YY1, ezh2, ikaros) (Fuxa et al. 2004; Liu et al. 2007; Reynaud et al. 2008), we favor a model in which one or indeed a coordinated network of these factors direct co-association of distal V genes with the DJ gene segment, and specify a specialized transcription-dependent recombination factory, and this will be an important avenue for future investigation.

Third, there is accumulating evidence to support a novel architectural role for ncRNA transcripts in establishing and maintaining subnuclear structural compartments (Clemson et al. 2009; Sasaki et al. 2009; Sunwoo et al. 2009). Recent live-imaging studies have shown that both the act of transcription and the nuclear

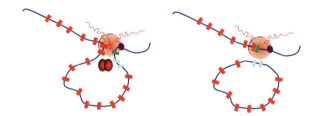

Fig. 3 Model of nuclear organization of Igh V to DJ recombination. *Orange circle*: RNA *Pol* II transcription factory; *red rectangles*: V genes; *dark green rectangle*: DJ recombined gene; *light green rectangles*: D genes; *purple oval*: Eu enhancer; *red ovals*: RAG1 and RAG2; *blue line*: 'looped' conformation of DNA sequence. One loop is depicted for simplicity, but as described in the text, there may be multiple loops; *red squiggly lines*: RNA transcripts. The diagram on the *left* depicts part of the V region and the recombined *DJ* gene segment juxtaposed in a single transcription factory due to simultaneous transcription: intergenic transcription from the V region and DJ transcription. Continuous Iu transcription is depicted from Eu. *A V* gene and the DJ segment are now in close proximity and the RAG recombinase complex is depicted nearby. The diagram on the *right* depicts the possible aftermath of V to DJ recombination: the VDJ gene segment remains in the transcription factory for high frequency transcription, and an excision circle is generated

RNAs themselves are required for nucleation, formation, and maintenance of a variety of nuclear bodies including nuclear speckles, paraspeckles, and Cajal bodies, employing a variety of RNAs (Mao et al. 2011; Shevtsov and Dundr 2011). The widespread nature of this mechanism in the substructures studied thus far indicates that RNA-primed biogenesis of nuclear bodies is a general feature of nuclear organization, which is likely to play a role in transcription factories also. While a specific subnuclear compartment for recombination has not yet been identified, the common mechanism of generating ncRNA transcripts may hold the key to how distal parts of the locus are held together for recombination. Given the size of the AgR loci and the need for stable proximity to DJ, it has been suggested that ncRNAs may modulate the higher order folding of the Igh loci in separate chromatin domains to facilitate proximity of all parts of the V region to the DJ region (Jhunjhunwala et al. 2009), and an interesting challenge for the future will be to model dynamic patterns of ncRNA expression, in a similar manner to modeling of Igh DNA sequence movement (Jhunjhunwala et al. 2008). The possibility that Igh ncRNAs might have a structural role resonates with several studies that show that, in addition to well documented Matrix Attachment Regions (MARs) flanking Eu (Jenuwein et al. 1997), the Igh V region has an unusually high number of MARs. These MARs exhibit heterogeneous binding of matrix-binding proteins, including SATB2 and BRIGHT, associated with repression and activation respectively, and this heterogeneity has been proposed to contribute to unequal V gene usage (Goebel et al. 2002). MARs organize chromatin into topological loops by anchoring DNA to the non-histone proteins of the nuclear matrix. If ncRNAs nucleate structure, they may recruit matrix-binding proteins to facilitate attachment of Igh loops to the nuclear matrix. Conceptually, the model of a stable nuclear sub-structure holding loops of the Igh together for transcription and recombination, whether mediated by ncRNA or not, is an attractive one. Transcription over the RSS and recombination involving DNA double-strand breaks are likely to be mutually exclusive. Thus intergenic transcription is an ideal 'buddy system' since engagement of intergenic sequences in the transcription factory will bring the V genes to the factory, but not too close to interfere with recombination (Fig. 3), or if so only briefly, while it will keep the V gene close to the factory for a relatively long time, since the long intergenic transcripts will take a considerable time to be transcribed, or 'reeled in' to the factory (Papantonis et al. 2010).

Finally, we have focused on epigenetic mechanisms, particularly non-coding RNA transcription at the level of the Ig loci, but genome-wide analyses are now re-defining our understanding of genome-wide genome regulation, and now the scene is set to discover the role of ncRNAs in lymphocyte development as a whole. Just as dynamic changes in expression of key transcription factors originally separated B cells from macrophages, or indeed T cells, in the future dynamic expression of lineage-specific ncRNAs may hold the epigenetic key. Some exciting studies have already demonstrated that numerous dynamic and functional alterations in the non-coding RNA repertoire occur through T cell development (Pang et al. 2009), so one can envisage the future discovery of a B cell-specific non-coding RNA 'master regulator' akin to the ncRNA produced in the p53 DNA damage response (Huarte et al. 2010).

13 Concluding Summary

Here we have explored our current understanding of some of the epigenetic mechanisms that regulate V(D)J recombination of the Igh loci to generate a diverse immunoglobulin repertoire. We have shown that in addition to localized epigenetic changes at *Igh* gene segments, the Igh loci undergo large-scale, stage-specific non-coding RNA transcription and have discussed its potential function and mode of action. Moreover, we have explored putative functions of ncRNAs as traffickers, structural components, and enhancers of Igh recombination. An important goal for the future will be to verify regulatory elements recently found, and to discover novel elements that activate the Igh V region. We have depicted the dynamic movement of Igh loci necessary for epigenetic activation of V(D)J recombination, and have proposed how the nuclear organization of V(D)J recombination and transcription may be intricately intertwined. With the aid of emerging technologies and modeling methods, an important goal for the future will be to generate an integrated and dynamic epigenetic model of V(D)J recombination incorporating all of the aspects discussed.

Acknowledgments The authors thank all members of the Corcoran group and the Chromatin and Gene Expression Laboratory for helpful discussions. The Biotechnology and Biological Sciences Research Council, UK supported this work.

References

Abarrategui I, Krangel MS (2006) Regulation of T cell receptor-alpha gene recombination by transcription. Nat Immunol 7:1109–1115

Abarrategui I, Krangel MS (2007) Noncoding transcription controls downstream promoters to regulate T cell receptor alpha recombination. EMBO J 26:4380–4390

Abarrategui I, Krangel MS (2009) Germline transcription: a key regulator of accessibility and recombination. Adv Exp Med Biol 650:93–102

Afshar R, Pierce S, Bolland DJ, Corcoran A, Oltz EM (2006) Regulation of IgH gene assembly: role of the intronic enhancer and 5'DQ52 region in targeting DHJH recombination. J Immunol 176:2439–2447

Bacher CP, Guggiari M, Brors B, Augui S, Clerc P, Avner P, Eils R, Heard E (2006) Transient colocalization of X-inactivation centres accompanies the initiation of X-inactivation. Nat Cell Biol 8:293–299

Bates JG, Cado D, Nolla H, Schlissel MS (2007) Chromosomal position of a VH gene segment determines its activation and inactivation as a substrate for V(D)J recombination. J Exp Med 204:3247–3256

Bernstein BE, Kamal M, Lindblad-Toh K, Bekiranov S, Bailey DK, Huebert DJ, McMahon S, Karlsson EK, Kulbokas EJ III, Gingeras TR, Schreiber SL, Lander ES (2005) Genomic maps and comparative analysis of histone modifications in human and mouse. Cell 120:169–181

Bertolino E, Reddy K, Medina KL, Parganas E, Ihle J, Singh H (2005) Regulation of interleukin 7-dependent immunoglobulin heavy-chain variable gene rearrangements by transcription factor STAT5. Nat Immunol 6:836–843

Bolland DJ, Wood AL, Afshar R, Featherstone K, Oltz EM, Corcoran AE (2007) Antisense intergenic transcription precedes Igh D-to-J recombination and is controlled by the intronic enhancer Emu. Mol Cell Biol 27:5523–5533

Bolland DJ, Wood AL, Johnston CM, Bunting SF, Morgan G, Chakalova L, Fraser PJ, Corcoran AE (2004) Antisense intergenic transcription in V(D)J recombination. Nat Immunol 5:630–637

Carninci P, Kasukawa T, Katayama S, Gough J, Frith MC, Maeda N, Oyama R, Ravasi T, Lenhard B, Wells C, Kodzius R, Shimokawa K, Bajic VB, Brenner SE, Batalov S, Forrest AR, Zavolan M, Davis MJ, Wilming LG, Aidinis V, Allen JE, Ambesi-Impiombato A, Apweiler R, Aturaliya RN, Bailey TL, Bansal M, Baxter L, Beisel KW, Bersano T, Bono H, Chalk AM, Chiu KP, Choudhary V, Christoffels A, Clutterbuck DR, Crowe ML, Dalla E, Dalrymple BP, de Bono B, Della Gatta G, di Bernardo D, Down T, Engstrom P, Fagiolini M, Faulkner G, Fletcher CF, Fukushima T, Furuno M, Futaki S, Gariboldi M, Georgii-Hemming P, Gingeras TR, Gojobori T, Green RE, Gustincich S, Harbers M, Hayashi Y, Hensch TK, Hirokawa N, Hill D, Huminiecki L, Iacono M, Ikeo K, Iwama A, Ishikawa T, Jakt M, Kanapin A, Katoh M, Kawasawa Y, Kelso J, Kitamura H, Kitano H, Kollias G, Krishnan SP, Kruger A, Kummerfeld SK, Kurochkin IV, Lareau LF, Lazarevic D, Lipovich L, Liu J, Liuni S, McWilliam S, Madan Babu M, Madera M, Marchionni L, Matsuda H, Matsuzawa S, Miki H, Mignone F, Miyake S, Morris K, Mottagui-Tabar S, Mulder N, Nakano N, Nakauchi H, Ng P, Nilsson R, Nishiguchi S, Nishikawa S et al (2005) The transcriptional landscape of the mammalian genome. Science 309:1559–1563

Chakraborty T, Chowdhury D, Keyes A, Jani A, Subrahmanyam R, Ivanova I, Sen R (2007) Repeat organization and epigenetic regulation of the DH-Cmu domain of the immunoglobulin heavy-chain gene locus. Mol Cell 27:842–850

Chakraborty T, Perlot T, Subrahmanyam R, Jani A, Goff PH, Zhang Y, Ivanova I, Alt FW, Sen R (2009) A 220-nucleotide deletion of the intronic enhancer reveals an epigenetic hierarchy in immunoglobulin heavy chain locus activation. J Exp Med 206:1019–1027

Chowdhury D, Sen R (2001) Stepwise activation of the immunoglobulin mu heavy chain gene locus. Embo J 20:6394–6403

Chowdhury D, Sen R (2003) Transient IL-7/IL-7R signaling provides a mechanism for feedback inhibition of immunoglobulin heavy chain gene rearrangements. Immunity 18:229–241

Clemson CM, Hutchinson JN, Sara SA, Ensminger AW, Fox AH, Chess A, Lawrence JB (2009) An architectural role for a nuclear noncoding RNA: NEAT1 RNA is essential for the structure of paraspeckles. Mol Cell 33:717–726

Corcoran AE (2005) Immunoglobulin locus silencing and allelic exclusion. Semin Immunol 17:141–154

Corcoran AE, Riddell A, Krooshoop D, Venkitaraman AR (1998) Impaired immunoglobulin gene rearrangement in mice lacking the IL-7 receptor. Nature 391:904–907

Cremer T, Cremer M (2010) Chromosome territories. Cold Spring Harb Perspect Biol 2:a003889

Degner SC, Wong TP, Jankevicius G, Feeney AJ (2009) Cutting edge: developmental stage-specific recruitment of cohesin to CTCF sites throughout immunoglobulin loci during B lymphocyte development. J Immunol 182:44–48

Ebert A, McManus S, Tagoh H, Medvedovic J, Salvagiotto G, Novatchkova M, Tamir I, Sommer A, Jaritz M, Busslinger M (2011) The distal Vh gene cluster of the Igh locus contains disticnt regulatory elements with Pax5 transcription factor-dependent activity in pro-B cells. Immunity 34:175–187

Engler P, Storb U (1999) Hypomethylation is necessary but not sufficient for V(D)J recombination within a transgenic substrate. Mol Immunol 36:1169–1173

Featherstone K, Wood AL, Bowen AJ, Corcoran AE (2010) The mouse immunoglobulin heavy chain V–D intergenic sequence contains insulators that may regulate ordered V(D)J recombination. J Biol Chem 285:9327–9338

Fraser P (2006) Transcriptional control thrown for a loop. Curr Opin Genet Dev 16:490–495

Fuxa M, Skok J, Souabni A, Salvagiotto G, Roldan E, Busslinger M (2004) Pax5 induces V-to-DJ rearrangements and locus contraction of the immunoglobulin heavy-chain gene. Genes Dev 18:411–422

Garrett FE, Emelyanov AV, Sepulveda MA, Flanagan P, Volpi S, Li F, Loukinov D, Eckhardt LA, Lobanenkov VV, Birshtein BK (2005) Chromatin architecture near a potential $3'$ end of the igh locus involves modular regulation of histone modifications during B cell development and in vivo occupancy at CTCF sites. Mol Cell Biol 25:1511–1525

Giallourakis CC, Franklin A, Guo C, Cheng HL, Yoon HS, Gallagher M, Perlot T, Andzelm M, Murphy AJ, Macdonald LE, Yancopoulos GD, Alt FW (2010) Elements between the IgH variable (V) and diversity (D) clusters influence antisense transcription and lineage-specific V(D)J recombination. Proc Natl Acad Sci USA 107:22207–22212

Goebel P, Montalbano A, Ayers N, Kompfner E, Dickinson L, Webb CF, Feeney AJ (2002) High frequency of matrix attachment regions and cut-like protein x/CCAAT-displacement protein and B cell regulator of IgH transcription binding sites flanking Ig V region genes. J Immunol 169:2477–2487

Grazini U, Zanardi F, Citterio E, Casola S, Goding CR, Mcblane F (2010) The RING domain of RAG1 ubiquitylates histone H3: a novel activity in chromatin-mediated regulation of V(D)J joining. Mol Cell 37:282–293

Gribnau J, Diderich K, Pruzina S, Calzolari R, Fraser P (2000) Intergenic transcription and developmental remodeling of chromatin subdomains in the human beta-globin locus. Mol Cell 5:377–386

Guttman M, Amit I, Garber M, French C, Lin M, Feldser D, Huarte M, Zuk O, Carey B, Cassady J, Cabili M, Jaenisch R, Mikkelsen T, Jacks T, Hacohen N, Bernstein B, Kellis M, Regev A, Rinn J, Lander E (2009) Chromatin signature reveals over a thousand highly conserved large non-coding RNAs in mammals. Nature 458:223–227

He Y, Vogelstein B, Velculescu VE, Papadopoulos N, Kinzler KW (2008) The antisense transcriptomes of human cells. Science 322:1855–1857

Hewitt SL, Yin B, Ji Y, Chaumeil J, Marszalek K, Tenthorey J, Salvagiotto G, Steinel N, Ramsey LB, Ghysdael J, Farrar MA, Sleckman BP, Schatz DG, Busslinger M, Bassing CH, Skok JA (2009) RAG-1 and ATM coordinate monoallelic recombination and nuclear positioning of immunoglobulin loci. Nat Immunol 10:655–664

Ho Y, Elefant F, Liebhaber SA, Cooke NE (2006) Locus control region transcription plays an active role in long-range gene activation. Mol Cell 23:365–375

Holliday R (1990) Mechanisms for the control of gene activity during development. Biol Rev Camb Philos Soc 65:431–471

Huarte M, Guttman M, Feldser D, Garber M, Koziol MJ, Kenzelmann-Broz D, Khalil AM, Zuk O, Amit I, Rabani M, Attardi LD, Regev A, Lander ES, Jacks T, Rinn JL (2010) A large intergenic noncoding RNA induced by p53 mediates global gene repression in the p53 response. Cell 142:409–419

Jenuwein T, Forrester WC, Fernandez-Herrero LA, Laible G, Dull M, Grosschedl R (1997) Extension of chromatin accessibility by nuclear matrix attachment regions. Nature 385:269–272

Jhunjhunwala S, van Zelm MC, Peak MM, Cutchin S, Riblet R, van Dongen JJ, Grosveld FG, Knoch TA, Murre C (2008) The 3D structure of the immunoglobulin heavy-chain locus: implications for long-range genomic interactions. Cell 133:265–279

Jhunjhunwala S, van Zelm MC, Peak MM, Murre C (2009) Chromatin architecture and the generation of antigen receptor diversity. Cell 138:435–448

Ji Y, Little AJ, Banerjee JK, Hao B, Oltz EM, Krangel MS, Schatz DG (2010a) Promoters, enhancers, and transcription target RAG1 binding during V(D)J recombination. J Exp Med 207:2809–2816

Ji Y, Resch W, Corbett E, Yamane A, Casellas R, Schatz DG (2010b) The in vivo pattern of binding of RAG1 and RAG2 to antigen receptor loci. Cell 141:419–431

Johnson K, Angelin-Duclos C, Park S, Calame KL (2003) Changes in histone acetylation are associated with differences in accessibility of V(H) gene segments to V–DJ recombination during B cell ontogeny and development. Mol Cell Biol 23:2438–2450

Johnson K, Pflugh DL, Yu D, Hesslein DG, Lin KI, Bothwell AL, Thomas-Tikhonenko A, Schatz DG, Calame K (2004) B cell-specific loss of histone 3 lysine 9 methylation in the V(H) locus depends on Pax5. Nat Immunol 5:853–861

Johnston CM, Wood AL, Bolland DJ, Corcoran AE (2006) Complete sequence assembly and characterization of the C57BL/6 mouse Ig heavy chain V region. J Immunol 176:4221–4234

Jones JM, Gellert M (2002) Ordered assembly of the V(D)J synaptic complex ensures accurate recombination. EMBO J 21:4162–4171

Kapranov P, Cheng J, Dike S, Nix DA, Duttagupta R, Willingham AT, Stadler PF, Hertel J, Hackermuller J, Hofacker IL, Bell I, Cheung E, Drenkow J, Dumais E, Patel S, Helt G, Ganesh M, Ghosh S, Piccolboni A, Sementchenko V, Tammana H, Gingeras TR (2007) RNA maps reveal new RNA classes and a possible function for pervasive transcription. Science 316:1484–1488

Katayama S, Tomaru Y, Kasukawa T, Waki K, Nakanishi M, Nakamura M, Nishida H, Yap CC, Suzuki M, Kawai J, Suzuki H, Carninci P, Hayashizaki Y, Wells C, Frith M, Ravasi T, Pang KC, Hallinan J, Mattick J, Hume DA, Lipovich L, Batalov S, Engstrom PG, Mizuno Y, Faghihi MA, Sandelin A, Chalk AM, Mottagui-Tabar S, Liang Z, Lenhard B, Wahlestedt C (2005) Antisense transcription in the mammalian transcriptome. Science 309:1564–1566

Khalil AM, Guttman M, Huarte M, Garber M, Raj A, Rivea Morales D, Thomas K, Presser A, Bernstein BE, van Oudenaarden A, Regev A, Lander ES, Rinn JL (2009) Many human large intergenic noncoding RNAs associate with chromatin-modifying complexes and affect gene expression. Proc Natl Acad Sci USA 106:11667–11672

Kim TK, Hemberg M, Gray JM, Costa AM, Bear DM, Wu J, Harmin DA, Laptewicz M, Barbara-Haley K, Kuersten S, Markenscoff-Papadimitriou E, Kuhl D, Bito H, Worley PF, Kreiman G, Greenberg ME (2010) Widespread transcription at neuronal activity-regulated enhancers. Nature 465:182–187

Kosak ST, Skok JA, Medina KL, Riblet R, Le Beau MM, Fisher AG, Singh H (2002) Subnuclear compartmentalization of immunoglobulin loci during lymphocyte development. Science 296:158–162

Koziol MJ, Rinn JL (2010) RNA traffic control of chromatin complexes. Curr Opin Genet Dev 20:142–148

Lennon GG, Perry RP (1985) C mu-containing transcripts initiate heterogeneously within the IgH enhancer region and contain a novel 5′-nontranslatable exon. Nature 318:475–478

Lieberman-Aiden E, van Berkum NL, Williams L, Imakaev M, Ragoczy T, Telling A, Amit I, Lajoie BR, Sabo PJ, Dorschner MO, Sandstrom R, Bernstein B, Bender MA, Groudine M, Gnirke A, Stamatoyannopoulos J, Mirny LA, Lander ES, Dekker J (2009) Comprehensive mapping of long-range interactions reveals folding principles of the human genome. Science 326:289–293

Liu H, Schmidt-Supprian M, Shi Y, Hobeika E, Barteneva N, Jumaa H, Pelanda R, Reth M, Skok J, Rajewsky K (2007) Yin Yang 1 is a critical regulator of B cell development. Genes Dev 21:1179–1189

Lucas JS, Bossen C, Murre C (2011) Transcription and recombination factories: common features? Curr Opin Cell Biol 23:318–324

Maes J, Chappaz S, Cavelier P, O'Neill L, Turner B, Rougeon F, Goodhardt M (2006) Activation of V(D)J recombination at the IgH chain JH locus occurs within a 6-kilobase chromatin domain and is associated with nucleosomal remodeling. J Immunol 176:5409–5417

Maes J, O'Neill LP, Cavelier P, Turner BM, Rougeon F, Goodhardt M (2001) Chromatin remodeling at the Ig loci prior to V(D)J recombination. J Immunol 167:866–874

Malin S, Mcmanus S, Cobaleda C, Novatchkova M, Delogu A, Bouillet P, Strasser A, Busslinger M (2010) Role of STAT5 in controlling cell survival and immunoglobulin gene recombination during pro-B cell development. Nat Immunol 11:171–179

Mao YS, Sunwoo H, Zhang B, Spector DL (2011) Direct visualization of the co-transcriptional assembly of a nuclear body by noncoding RNAs. Nat Cell Biol 13:95–101

Masternak K, Peyraud N, Krawczyk M, Barras E, Reith W (2003) Chromatin remodeling and extragenic transcription at the MHC class II locus control region. Nat Immunol 4:132–137

Matthews AG, Kuo AJ, Ramon-Maiques S, Han S, Champagne KS, Ivanov D, Gallardo M, Carney D, Cheung P, Ciccone DN, Walter KL, Utz PJ, Shi Y, Kutateladze TG, Yang W, Gozani O, Oettinger MA (2007) RAG2 PHD finger couples histone H3 lysine 4 trimethylation with V(D)J recombination. Nature 450:1106–1110

Mercer TR, Dinger ME, Mattick JS (2009) Long non-coding RNAs: insights into functions. Nat Rev Genet 10:155–159

Mitchell JA, Fraser P (2008) Transcription factories are nuclear subcompartments that remain in the absence of transcription. Genes Dev 22:20–25

Mito Y, Henikoff JG, Henikoff S (2005) Genome-scale profiling of histone H3.3 replacement patterns. Nat Genet 37:1090–1097

Morshead KB, Ciccone DN, Taverna SD, Allis CD, Oettinger MA (2003) Antigen receptor loci poised for V(D)J rearrangement are broadly associated with BRG1 and flanked by peaks of histone H3 dimethylated at lysine 4. Proc Natl Acad Sci USA 100:11577–11582

Mostoslavsky R, Singh N, Kirillov A, Pelanda R, Cedar H, Chess A, Bergman Y (1998) Kappa chain monoallelic demethylation and the establishment of allelic exclusion. Genes Dev 12:1801–1811

Nagano T, Mitchell JA, Sanz LA, Pauler FM, Ferguson-Smith AC, Feil R, Fraser P (2008) The Air noncoding RNA epigenetically silences transcription by targeting G9a to chromatin. Science 322:1717–1720

Ng HH, Robert F, Young RA, Struhl K (2003) Targeted recruitment of Set1 histone methylase by elongating Pol II provides a localized mark and memory of recent transcriptional activity. Mol Cell 11:709–719

Nitschke L, Kestler J, Tallone T, Pelkonen S, Pelkonen J (2001) Deletion of the DQ52 element within the Ig heavy chain locus leads to a selective reduction in VDJ recombination and altered D gene usage. J Immunol 166:2540–2552

Orom UA, Derrien T, Beringer M, Gumireddy K, Gardini A, Bussotti G, Lai F, Zytnicki M, Notredame C, Huang Q, Guigo R, Shiekhattar R (2010) Long noncoding RNAs with enhancer-like function in human cells. Cell 143:46–58

Orphanides G, Reinberg D (2000) RNA polymerase II elongation through chromatin. Nature 407:471–475

Osborne CS, Chakalova L, Brown KE, Carter D, Horton A, Debrand E, Goyenechea B, Mitchell JA, Lopes S, Reik W, Fraser P (2004) Active genes dynamically colocalize to shared sites of ongoing transcription. Nat Genet 36:1065–1071

Osborne CS, Chakalova L, Mitchell JA, Horton A, Wood AL, Bolland DJ, Corcoran AE, Fraser P (2007) Myc dynamically and preferentially relocates to a transcription factory occupied by Igh. PLoS Biol 5:e192

Osipovich OA, Subrahmanyam R, Pierce S, Sen R, Oltz EM (2009) Cutting edge: SWI/SNF mediates antisense Igh transcription and locus-wide accessibility in B cell precursors. J Immunol 183:1509–1513

Pang KC, Dinger ME, Mercer TR, Malquori L, Grimmond SM, Chen W, Mattick JS (2009) Genome-wide identification of long noncoding RNAs in CD8 + T cells. J Immunol 182:7738–7748

Papantonis A, Larkin JD, Wada Y, Ohta Y, Ihara S, Kodama T, Cook PR (2010) Active RNA polymerases: mobile or immobile molecular machines? PLoS Biol 8:e1000419

Pawlitzky I, Angeles CV, Siegel AM, Stanton ML, Riblet R, Brodeur PH (2006) Identification of a candidate regulatory element within the 5′ flanking region of the mouse Igh locus defined by pro-B cell-specific hypersensitivity associated with binding of PU.1, Pax5, and E2A. J Immunol 176:6839–6851

Perlot T, Alt FW, Bassing CH, Suh H, Pinaud E (2005) Elucidation of IgH intronic enhancer functions via germ-line deletion. Proc Natl Acad Sci USA 102:14362–14367

Perlot T, Pawlitzky I, Manis JP, Zarrin AA, Brodeur PH, Alt FW (2010) Analysis of mice lacking DNaseI hypersensitive sites at the 5′ end of the IgH locus. PLoS One 5:e13992

Phillips JE, Corces VG (2009) CTCF: master weaver of the genome. Cell 137:1194–1211

Ponting CP, Oliver PL, Reik W (2009) Evolution and functions of long noncoding RNAs. Cell 136:629–641

Ragoczy T, Bender MA, Telling A, Byron R, Groudine M (2006) The locus control region is required for association of the murine beta-globin locus with engaged transcription factories during erythroid maturation. Genes Dev 20:1447–1457

Reddy KL, Zullo JM, Bertolino E, Singh H (2008) Transcriptional repression mediated by repositioning of genes to the nuclear lamina. Nature 452:243–247

Reth MG, Alt FW (1984) Novel immunoglobulin heavy chains are produced from DJH gene segment rearrangements in lymphoid cells. Nature 312:418–423

Reynaud D, Demarco IA, Reddy KL, Schjerven H, Bertolino E, Chen Z, Smale ST, Winandy S, Singh H (2008) Regulation of B cell fate commitment and immunoglobulin heavy-chain gene rearrangements by Ikaros. Nat Immunol 9:927–936

Rinn JL, Kertesz M, Wang JK, Squazzo SL, Xu X, Brugmann SA, Goodnough LH, Helms JA, Farnham PJ, Segal E, Chang HY (2007) Functional Demarcation of Active and Silent Chromatin Domains in Human HOX Loci by Noncoding RNAs. Cell 129:1311–1323

Roldan E, Fuxa M, Chong W, Martinez D, Novatchkova M, Busslinger M, Skok JA (2005) Locus 'decontraction' and centromeric recruitment contribute to allelic exclusion of the immuno-globulin heavy-chain gene. Nat Immunol 6:31–41

Sasaki YT, Ideue T, Sano M, Mituyama T, Hirose T (2009) MENepsilon/beta noncoding RNAs are essential for structural integrity of nuclear paraspeckles. Proc Natl Acad Sci U S A 106:2525–2530

Sayegh C, Jhunjhunwala S, Riblet R, Murre C (2005) Visualization of looping involving the immunoglobulin heavy-chain locus in developing B cells. Genes Dev 19:322–327

Schmitt S, Prestel M, Paro R (2005) Intergenic transcription through a polycomb group response element counteracts silencing. Genes Dev 19:697–708

Schoenfelder S, Clay I, Fraser P (2010a) The transcriptional interactome: gene expression in 3D. Curr Opin Genet Dev 20:127–133

Schoenfelder S, Sexton T, Chakalova L, Cope NF, Horton A, Andrews S, Kurukuti S, Mitchell JA, Umlauf D, Dimitrova DS, Eskiw CH, Luo Y, Wei CL, Ruan Y, Bieker JJ, Fraser P (2010b) Preferential associations between co-regulated genes reveal a transcriptional interactome in erythroid cells. Nat Genet 42:53–61

Sessa L, Breiling A, Lavorgna G, Silvestri L, Casari G, Orlando V (2007) Noncoding RNA synthesis and loss of Polycomb group repression accompanies the colinear activation of the human HOXA cluster. RNA 13:223–239

Shevtsov SP, Dundr M (2011) Nucleation of nuclear bodies by RNA. Nat Cell Biol 13:167–173

Sleutels F, Zwart R, Barlow DP (2002) The non-coding Air RNA is required for silencing autosomal imprinted genes. Nature 415:810–813

Spector DL (2003) The dynamics of chromosome organization and gene regulation. Annu Rev Biochem 72:573–608

Stanhope-Baker P, Hudson KM, Shaffer AL, Constantinescu A, Schlissel MS (1996) Cell type-specific chromatin structure determines the targeting of V(D)J recombinase activity in vitro. Cell 85:887–897

Storb U, Arp B (1983) Methylation patterns of immunoglobulin genes in lymphoid cells: correlation of expression and differentiation with undermethylation. Proc Natl Acad Sci USA 80:6642–6646

Su IH, Basavaraj A, Krutchinsky AN, Hobert O, Ullrich A, Chait BT, Tarakhovsky A (2003) Ezh2 controls B cell development through histone H3 methylation and Igh rearrangement. Nat Immunol 4:124–131

Sunwoo H, Dinger ME, Wilusz JE, Amaral PP, Mattick JS, Spector DL (2009) MEN epsilon/beta nuclear-retained non-coding RNAs are up-regulated upon muscle differentiation and are essential components of paraspeckles. Genome Res 19:347–359

Thompson A, Timmers E, Schuurman RK, Hendriks RW (1995) Immunoglobulin heavy chain germ-line JH-C mu transcription in human precursor B lymphocytes initiates in a unique region upstream of DQ52. Eur J Immunol 25:257–261

Tsai MC, Manor O, Wan Y, Mosammaparast N, Wang JK, Lan F, Shi Y, Segal E, Chang HY (2010) Long noncoding RNA as modular scaffold of histone modification complexes. Science 329: 689–693

Volpi EV, Chevret E, Jones T, Vatcheva R, Williamson J, Beck S, Campbell RD, Goldsworthy M, Powis SH, Ragoussis J, Trowsdale J, Sheer D (2000) Large-scale chromatin organization of the major histocompatibility complex and other regions of human chromosome 6 and its response to interferon in interphase nuclei. J Cell Sci 113(Pt 9):1565–1576

Wilson CJ, Chao DM, Imbalzano AN, Schnitzler GR, Kingston RE, Young RA (1996) RNA polymerase II holoenzyme contains SWI/SNF regulators involved in chromatin remodeling. Cell 84:235–244

Xu C-R, Feeney AJ (2009) The epigenetic profile of Ig genes is dynamically regulated during B cell differentiation and is modulated by pre-B cell receptor signaling. J Immunol 182:1362–1369

Xu CR, Schaffer L, Head SR, Feeney AJ (2008) Reciprocal patterns of methylation of H3K36 and H3K27 on proximal vs. distal IgVH genes are modulated by IL-7 and Pax5. Proc Natl Acad Sci USA 105:8685–8690

Yancopoulos GD, Alt FW (1985) Developmentally controlled and tissue-specific expression of unrearranged VH gene segments. Cell 40:271–281

Yang Q, Riblet R, Schildkraut CL (2005) Sites that direct nuclear compartmentalization are near the 5' end of the mouse immunoglobulin heavy-chain locus. Mol Cell Biol 25:6021–6030

Zhao J, Sun BK, Erwin JA, Song JJ, Lee JT (2008) Polycomb proteins targeted by a short repeat RNA to the mouse X chromosome. Science 322:750–756

Genetic and Epigenetic Regulation of *Tcrb* Gene Assembly

Michael L. Sikes and Eugene M. Oltz

Abstract Vertebrate development requires the formation of multiple cell types from a single genetic blueprint, an extraordinary feat that is guided by the dynamic and finely tuned reprogramming of gene expression. The sophisticated orchestration of gene expression programs is driven primarily by changes in the patterns of covalent chromatin modifications. These epigenetic changes are directed by cis elements, positioned across the genome, which provide docking sites for transcription factors and associated chromatin modifiers. Epigenetic changes impact all aspects of gene regulation, governing association with the machinery that drives transcription, replication, repair and recombination, a regulatory relationship that is dramatically illustrated in developing lymphocytes. The program of somatic rearrangements that assemble antigen receptor genes in precursor B and T cells has proven to be a fertile system for elucidating relationships between the genetic and epigenetic components of gene regulation. This chapter describes our current understanding of the cross-talk between key genetic elements and epigenetic programs during recombination of the *Tcrb* locus in developing T cells, how each contributes to the regulation of chromatin accessibility at individual DNA targets for recombination, and potential mechanisms that coordinate their actions.

M. L. Sikes (✉)
Department of Microbiology, North Carolina State University, 100 Derieux Place,
Campus Box 7615, Raleigh, NC 27695, USA
e-mail: mlsikes@ncsu.edu

E. M. Oltz
Department of Pathology and Immunology, Washington University School of Medicine,
660 Euclid Ave., Campus Box 8118, St. Louis, MO 63110, USA
e-mail: eoltz@pathology.wustl.edu

Current Topics in Microbiology and Immunology (2012) 356: 91–116
DOI: 10.1007/82_2011_138
© Springer-Verlag Berlin Heidelberg 2011
Published Online: 18 June 2011

Abbreviations

RSS	Recombination signal sequence
Tcrb	T cell receptor β locus
Tcra	T cell receptor α locus
DN	Double negative
DP	Double positive
Eβ	*Tcrb* enhancer
PDβ1	Dβ1 promoter
ChIP	Chromatin immunoprecipitation
Rag	Recombination activating gene
ACE	Accessibility control element

Contents

1 Introduction 92
2 *Tcr* Gene Assembly During Thymocyte Development 93
3 Genetic Determinants of Recombination Efficiency 95
    3.1 RSSs and Ordered *Tcrb* Assembly 95
    3.2 Transcriptional Promoters and Enhancers 96
4 Epigenetic Determinants: Regulating RSS Chromatin Accessibility 98
    4.1 The Epigenetic Landscape 98
    4.2 Promoter-mediated DβJβ Recombination 99
    4.3 Promoter-Mediated Accessibility of Vβ-to-DβJβ Recombination 103
5 Regulating Locus Assembly Beyond 12/23 103
6 Long-range Changes in Locus Conformation 105
7 *Tcrb* Allelic Exclusion 107
8 Conclusions 110
References 111

1 Introduction

The B and T cell antigen receptors present an enormously diverse and highly adaptable repertoire to protect us from an ever-evolving world of pathogens. Receptor diversity is achieved via the random assembly of exons encoding hypervariable domains for immunoglobulin (Ig) and T cell receptor (*Tcr*) genes in precursor B and T lymphocytes, respectively (Feeney 2009; Krangel 2009). The assembly process consists of a tightly regulated series of genomic rearrangements, termed V(D)J recombination, which refers to the variable, diversity, and joining gene segments that are targeted for recombination in each precursor cell. A functional antigen-binding domain requires that one V segment, selected from a large pool, ligate to one of several downstream J segments; or in loci that contain D elements, one V must be joined to one D following its rearrangement to a selected J element. The enzymatic mechanisms that govern V(D)J recombination are conserved between B and T cells (Schatz and Spanopoulou 2005). Generally speaking, recombination proceeds via: (1) binding of lymphocyte-specific RAG-1 and RAG-2 proteins to recombination signal sequences (RSSs) that flank all gene segments, (2) assembly of a synaptic recombination complex that includes the recombinase and two compatible gene segments, (3) introduction of double-strand DNA breaks between synaptic RSSs and their associated coding segments, (4) coding end modification, and (5) coding joint formation via ubiquitous DNA repair pathways.

Although the various recombination events that occur within antigen receptor loci utilize the same enzymatic machinery, individual gene segments differ dramatically in their competency and availability for rearrangement. These important features govern the efficiency of recombination and are controlled by cell lineage- and developmental stage-specific mechanisms (Thomas et al. 2009). A major component of these regulatory mechanisms is the modulation of chromatin, a macromolecular complex of DNA and histone octamers, which can impede RAG binding and cleavage of RSSs. Noting that unrearranged Igh gene segments are transcribed in a developmental- and lineage-specific pattern that parallels their recombination, Yancopoulos and Alt (1985) proposed that differential control of chromatin accessibility at gene segments may account for the specificity of recombination (Yancopoulos and Alt 1985).

Launching from this seminal finding, the question of how gene segments are rendered accessible or inaccessible to recombinase has been a subject of intense research over the intervening 25 years. A wealth of data correlates recombinase accessibility with other epigenetic marks of chromatin opening, and has implicated a number of genetic elements including enhancers, promoters, and specific RSSs as determinants of recombination efficiency (Osipovich and Oltz 2010). Nonetheless, the spectrum of mechanisms employed by precursor lymphocytes to regulate RSS accessibility has remained stubbornly elusive. In this chapter, we describe our current understanding of accessibility control mechanisms used by thymocytes to assemble functional *Tcrb* genes, with a focus on the extensive crosstalk that occurs between genetic elements and epigenetic pathways.

2 *Tcr* Gene Assembly During Thymocyte Development

B and T cells develop from a bone marrow-derived common lymphoid progenitor (Murre 2009). Extrathymic progenitors migrate from the bone marrow to the subcapsular region of the thymic cortex, where they differentiate into early thymocyte progenitors, or ETPs (formerly CD4$^-$CD8$^-$ DN1), a process that requires induction of several key genes including c-kit, CD44 and RAG-1/2 (Porritt et al. 2004). Early thymocyte development proceeds along a defined pathway beginning with CD44$^+$CD25$^-$ ETPs, which then commit to the T cell lineage at the CD44$^+$CD25$^+$ DN2 stage, followed by the CD44$^-$CD25$^+$ DN3 and

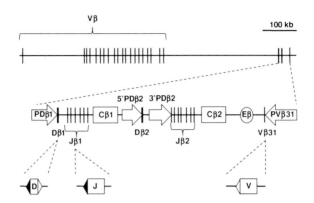

Fig. 1 Genetic organization of the mouse *Tcrb* locus. The 3′ portion of the locus containing the two DJCβ gene segment clusters and Vβ14 is enlarged in the middle panel. The lone Dβ1 promoter and dual Dβ2 promoters are shown (*blockarrows*) along with the Eβ enhancer (*circle*). RSS organization of *Tcrb* is shown in the *lower panel*. 12-RSSs (*filled triangles*) are positioned 5′ of the D and J segments, and 23-RSSs (*opentriangles*) are positioned 3′ of the D and V segments (Vβ31 is inverted)

CD44⁻CD25⁻ DN4 stages (Hayday and Pennington 2007). Although ETPs and DN2 cells retain limited proliferative capacity, a daily influx of extrathymic progenitors is required to sustain T cell development. The cell cycle is arrested in DN3/4 thymocytes, the major populations in which V(D)J recombinase completes assembly of the *Tcrb*, *Tcrd* and *Tcrg* loci. Generation of in-frame joints on both the *Tcrd*, and *Tcrg* loci leads to expression of a complete γδ TCR on the cell surface, cessation of further V(D)J recombination, and emigration of a new γδ T cell from the thymus. However, γδ cells ultimately account for <10% of circulating T lymphocytes. Rather, the majority of DN thymocytes complete assembly of *Tcrb* before they express a functional γδ TCR (Hayday and Pennington 2007).

The mouse *Tcrb* locus encompasses ∼700 kb of chromosome 6 (Fig. 1). The vast majority of the locus (∼624 kb) is dedicated to 31 functional Vβ elements and two arrays of trypsinogen genes that flank the bulk of Vβ elements. The 3′ portion of the locus (∼40 kb) contains two separate DJβCβ clusters, each composed of a single Dβ element positioned 5′ of six functional Jβ elements and a constant region. Downstream of the DJβCβ clusters, the locus contains the only known enhancer, Eβ, and a single inverted Vβ (Vβ31, formerly Vβ14). DN thymocytes assemble *Tcrb* in two steps, initiating with D-to-J recombination on both alleles followed by mono-allelic rearrangement of a DJ joint with one of the 31 Vβ segments (Krangel 2009). This staged assembly process is intimately linked to thymocyte development; both DJβ cassettes initiate recombination in late ETP/DN2 cells, while Vβ elements do not become recombinationally active until the DN3 stage of development (Livak 2004).

Although only one in three V-to-DJβ rearrangements are expected to maintain a proper translational frame, serial organization of the two DJβ cassettes on each *Tcrb* allele provides individual thymocytes with as many as four opportunities for successful *Tcrb* gene assembly. Completion of *Tcrb* recombination in DN3 cells serves as an obligate checkpoint for continued thymocyte development. So-called β selection is initiated by the pre-TCR, a signaling complex consisting of the newly expressed TCRβ protein, CD3 co-receptors, and the surrogate TCRα protein, called pTα. Expression of pre-TCR on a DN4 thymocyte triggers a series of irreversible events that include the cessation of additional *Tcrb* recombination (allelic exclusion), up-regulation of CD4/CD8, and a proliferative burst that is closely followed by *Tcra* recombination in the $CD4^+$ $CD8^+$ double positive (DP) progeny (Brady et al. 2010). Allelic exclusion at *Tcrb* does not prevent continued assembly of *Tcrd* and *Tcrg* alleles in DP cells. However, because *Tcrd* is positioned within the *Tcra* locus, activation of Vα-Jα recombination deletes *Tcrd* and dramatically limits the production of $\gamma\delta$ cells (Krangel 2009).

Analogous rearrangement and development patterns are observed for the Ig heavy and then light chain loci during B cell development (Cobb et al. 2006). As with *Tcrb* in DN thymocytes, Igh genes are assembled in bone marrow $B220^+$ $CD43^+$ pro-B cells via a two-step process in which D-to-J is followed by V-to-DJ recombination. Assembly of an in-frame $V(D)J_H$ joint leads to a burst of proliferation and developmental progression to pre-B cells where the Igk and Igl light chain genes are assembled.

The ordered nature of recombination in developing lymphocytes requires recombinase to be directed initially to one set of gene segments, and then redirected to a second set of gene segments within the appropriate developmental windows. If recombination is controlled at the level of gene segment accessibility to recombinase, how then is accessibility tethered to lymphocyte development? The chromatin structure that characterizes eukaryotic DNA can provide a significant barrier to RAG-1/2 binding at RSS targets (Kwon et al. 2000; Stanhope-Baker et al. 1996), and a variety of studies have shown that transcriptional promoters, enhancers, and other regulatory elements alter chromatin structure and impact recombinational accessibility (Osipovich and Oltz 2010). Using *Tcrb* as a model to ultimately understand the epigenetic changes in chromatin structure that regulate antigen receptor gene accessibility, we begin with a discussion of the genetic elements that coordinate *Tcrb* assembly.

3 Genetic Determinants of Recombination Efficiency

3.1 *RSSs and Ordered* Tcrb *Assembly*

All RSSs are composed of a conserved palindromic heptamer that abuts the V, D, or J coding sequences. Each heptamer is separated from an A/T-rich nonamer element by less conserved spacer sequences of either 12 or 23 bp in length. During recombination, the RSS is predominantly bound by the RAG-1 subunit of recombinase, which spans the spacer sequence to simultaneously contact both the nonamer and heptamer elements in a near-planar fashion (Swanson and Desiderio 1998). Indeed, the differential spacer lengths accommodate RAG-1 binding requirements by allowing just over one (12 bp) or two helical turns (23 bp) between the conserved elements. Deletion or addition of even a single base in the 12-RSS or 23-RSS spacer dramatically impair RAG-1 binding (Steen et al. 1997). RAG-2 functions primarily as a co-factor, reading the transcription-dependent deposition of histone H3 lysine 4 trimethylation (H3K4me3), and enhancing RAG-1 binding, presumably at H3K4me3-enriched RSSs (Matthews et al. 2007; Swanson et al. 2009).

The precise distribution of 12- versus 23-RSSs is consistent among gene segment types within a locus. For example, all $V\alpha$ gene segments are flanked by 23-RSSs, while all $J\alpha$ segments are flanked by 12-RSSs. Although the stoichiometry of recombinase remains undefined, multi-subunit complexes of RAG-1/2 likely assemble on a single available RSS, and then capture a second RSS to form a synaptic cleavage complex (Jones and Gellert 2002; Mundy et al. 2002). Synapses are only formed between two RSSs of different spacer lengths, a restriction known as the "12-23 rule" (Eastman et al. 1996), which prevents recombination between equivalent gene segments (e.g., V-to-V or J-to-J recombination). In the *Tcrb* and *Tcrd* loci, V elements are flanked by 23-RSSs, J elements by 12-RSSs, and the intermediate D elements are flanked 3' by 23-RSSs (for D-to-J recombination) and 5' by 12-RSSs (to accommodate V-to-DJ joining). In these loci, 12-23 restriction is insufficient to limit direct V-to-J or internal D-to-D joining (common in *Tcrd*, but infrequent in *Tcrb*). Rather, additional mechanisms act beyond the 12-23 rule (B12/23) to enforce $D\beta$ utilization (discussed in more detail below).

Consistent with the relatively short length of RSS elements and their simplified structural requirements, a number of cryptic RSS elements have been identified at hotspots for leukemic translocations (e.g., within c-*Myc*) (Jankovic et al. 2007), and many more have been predicted to reside across the human genome (Cowell et al. 2003). Indeed, recombinase-dependent translocations have been found in a large number of lymphoid cancers, fusing critical growth control genes with transcriptionally active antigen receptor genes (Jankovic et al. 2007). Nonetheless, inappropriate recombinase targeting is remarkably infrequent, again suggesting that the genetic determinants of recombination must operate within a larger regulatory framework that mediates their accessibility to recombinase.

3.2 Transcriptional Promoters and Enhancers

The complexity of regulatory mechanisms governing the assembly of antigen receptor loci is underscored by the requirement that one or more enhancers, usually positioned 3' of the V, D, J clusters, must communicate with a large collection of promoter elements splayed over great distances. Each V element is flanked by a 5' promoter that is responsible for driving expression of functionally rearranged Ig *or Tcr* genes, while separate promoters positioned within D or J gene segment clusters drive expression of sterile transcripts prior to locus assembly (Abarrategui and Krangel 2009).

In *Tcrb*, germ-line transcription across the two DβJβCβ clusters requires a lone 3′ enhancer (Eβ), which facilitates activation of promoters flanking each of the two Dβ segments (McMillan and Sikes 2008; Sikes et al. 1998). PDβ1, the first germ-line *Tcrb* promoter discovered, is positioned immediately 5′ of the Dβ1 12-RSS, and utilizes a TATA element situated within the RSS spacer to initiate transcription of Dβ1 coding sequences. PDβ1 binds a variety of ubiquitous and T cell-restricted transcription factors including SP1, GATA-3, and members of the ETS, RUNX and bHLH families, most of which also bind Eβ (Doty et al. 1999; Sikes et al. 1998; Tripathi et al. 2000). Mice lacking either Eβ or PDβ1 are dramatically impaired for *Tcrb* gene assembly. Indeed, Eβ-deficient mice recapitulate the total loss of *Tcrb* assembly seen in RAG-deficient animals (Bories et al. 1996; Bouvier et al. 1996), whereas targeted deletion of PDβ1 specifically attenuates DJβ1 rearrangement without altering DJβ2 or subsequent Vβ-to-(Dβ2)Jβ rearrangements (Whitehurst et al. 1999). These distinct phenotypes suggest that Eβ modulates chromatin accessibility across both DJβ clusters ($\sim$20 kb), while the contributions of individual promoters are much more localized.

Promoters controlling transcription within the DJβ2 cluster have been defined only recently, with separate elements situated upstream and downstream of the Dβ2 coding sequence. Analogous to PDβ1, the 5′PDβ2 promoter is positioned immediately upstream of Dβ2 and binds the GATA-3, RUNX-1 and E47 transcription factors (McMillan and Sikes 2009). However, in striking contrast to PDβ1, the 5′PDβ2 is inactive prior to Dβ2-to-Jβ2 recombination. Rather, germline transcription in the Dβ2Jβ2Cβ2 cassette is driven by an NFκB-dependent promoter located several hundred bp downstream of Dβ2 (3′PDβ2) (McMillan and Sikes 2008). Repression of 5′PDβ2 activity is mediated by the constitutively expressed bHLH protein, USF-1, which binds a target E-box in the spacer sequence of the Dβ2 12-RSS. Evidence in cell lines and thymocytes treated with UV radiation suggests that induction of the p38 MAPK, perhaps in response to recombination-dependent activation of DNA repair pathways, leads to loss of USF-1 binding at Dβ2 and the activation of 5′PDβ2 (McMillan and Sikes, submitted). Thus, in DN2 thymocytes, active transcriptional promoters are situated adjacent to Dβ1 but significantly downstream of Dβ2 prior to recombination. Following DβJβ recombination, 5′PDβ2 is activated and both DJβ1 and DJβ2 joints are transcribed as thymocytes and prepare for DJβ recombination with distant Vβ gene segments.

In contrast to Dβ promoters, relatively little is known about the developmental regulation of promoters that drive Vβ transcription. These promoters are responsible for germ-line and rearranged expression of associated Vβ elements, share a common cAMP response element-like decamer motif, and may also bind ETS and RUNX family proteins (Halle et al. 1997). It is likely that Vβ promoters also mediate access to recombinase in a manner similar to the Dβ promoters. Indeed, knockout of the Vβ14 (formerly Vβ13) promoter led to a tenfold decrease in Vβ14 rearrangement, while replacement of PVβ14 with the SV40 minimal promoter essentially restored wild-type levels and appropriate timing of Vβ14 recombination (Ryu et al. 2004). Unlike Dβ promoters, however, Vβ promoters do not appear to require Eβ for their transcriptional activation in DN3 cells (Mathieu et al. 2000). How *Tcrb* promoters and enhancers function as accessibility control elements (ACEs) has become an area of intense focus and will be discussed at length below.

4 Epigenetic Determinants: Regulating RSS Chromatin Accessibility

4.1 The Epigenetic Landscape

Genomic DNA is packaged in the eukaryotic nucleus as chromatin, a flexible and dynamic nucleoprotein structure in which DNA is wound around repeating cores of histone proteins (dimers of histones H2A, H2B, H3, and H4). In addition to compacting the genome, chromatin provides an epigenetic platform for regulating the access of DNA binding proteins (Jenuwein and Allis 2001; Khorasanizadeh 2004). Seminal studies have correlated covalent modifications, including CpG methylation, histone acetylation, and histone methylation, with a gene's state of chromatin condensation and, by extension, its accessibility (Osipovich and Oltz 2010). Condensed heterochromatic genes are silent, and tend to be enriched for hypermethylated CpG dinucleotides and for histone H3 that carries methyl tags on lysine 9 (H3K9me). Conversely, accessible genes are organized in relaxed euchromatin that is largely devoid of CpG or H3K9 methylation, and instead is enriched for acetylation of various H3 and H4 lysine residues (H3ac and H4ac) (Jenuwein and Allis 2001). The transcriptional status of a gene also correlates with the local patterns of histone modifications. Transcriptionally silent genes are enriched for H3K27 trimethylation (H3K27me3), whereas transcribed genes are enriched for H3K4me3 at their promoters (Hublitz et al. 2009; Lee et al. 2010). Together with a variety of other histone tags, these methylation marks provide a sensitive, multi-layered mechanism to independently regulate the activity of discrete chromatin regions in response to environmental, metabolic, or developmental signals.

The epigenetic organization of antigen receptor genes is a critical determinant of their recombinational accessibility. Like most DNA binding proteins, RAG-1 and -2 are extremely sensitive to the chromatin state of their DNA targets. In vitro chromatinization of double stranded RSS templates impairs RAG-mediated cleavage (Golding et al. 1999; Kwon et al. 2000; Stanhope-Baker et al. 1996). Antigen receptor loci in non-lymphoid cells exist in a facultative heterochromatic state marked by elevated levels of repressive methylation at CpG and H3K9 (Mathieu et al. 2000; Osipovich et al. 2004). RAG-1/2 proteins ectopically expressed in such cells fail to target endogenous antigen receptor genes, though they readily recombine RSSs in naked DNA substrates introduced by transient

Genetic and Epigenetic Regulation of *Tcrb* Gene Assembly

transfection (Schatz and Baltimore 1988). The epigenetic landscape of antigen receptor loci is strikingly different in developing lymphocytes. For example, *Tcrb* in DN thymocytes is largely devoid of CpG methylation (Chattopadhyay et al. 1998), and is enriched for H3/H4 acetylation (Ji et al. 2010a; Morshead et al. 2003; Tripathi et al. 2002) as well as foci of H3K4me3 (Ji et al. 2010b). Following *Tcrb* assembly and β selection, heterochromatic marks are restored across much of the remaining germ-line Vβ segments in DP cells, in which allelic exclusion is enforced (Tripathi et al. 2002).

Developmental changes in the epigenetic landscape of *Tcrb* are largely due to the actions of Eβ as well as promoters proximal to Vβ and Dβ elements. Loss of enhancer activity in Eβ-deficient mice converts the *Tcrb* epigenetic landscape from an open to a closed state in DN cells, leaving the DβJβCβ and Vβ31 regions hypermethylated at CpG dinucleotides and associated with histones bearing H3K9me rather than acetylation marks (Fig. 2) (Mathieu et al. 2000). In contrast to enhancer-dependent chromatin opening, deletion of PDβ1 has little impact on the histone acetylation within the DJβ1 cluster (Ji et al. 2010a; Sikes et al. 2002). Additionally, PDβ1 deletion does not seem to impair the general accessibility of Jβ gene segments but specifically impairs restriction endonuclease access at sites immediately proximal to Dβ1 (Oestreich et al. 2006). Indeed, direct analysis of nucleosome phasing in *Tcrb* shows that, even in the absence of PDβ1, the Dβ1 23-RSS and Jβ1.1 12-RSS remain essentially free of histones (Kondilis-Mangum et al. 2010). It is intriguing therefore that RAG-1 binding is dramatically attenuated across the entire DJβ1 cluster in PDβ1-deficient mice (Ji et al. 2010a). This apparent disconnect between *chromatin* accessibility and *recombinational* accessibility shows that while enhancer-dependent chromatin opening may be essential, it is not sufficient to direct recombinational accessibility in the absence of promoter activity.

4.2 Promoter-mediated DβJβ Recombination

The most obvious function of promoters with regard to recombinase accessibility is transcriptional activation, which induces a variety of chromatin alterations both upstream of and within transcribed sequences (Li et al. 2007). This general concept meshes well with the seminal observations that drove the accessibility model, i.e., the tight correlations between germ-line transcription and recombination of antigen receptor gene segments (Van Ness et al. 1981; Yancopoulos and Alt 1985). These links were bolstered by targeted enhancer deletions, which not only block recombination and opening of regional chromatin, but also germ-line transcription (Osipovich and Oltz 2010). But does transcription directly contribute to recombinational accessibility, or is it simply a by-product of promoter activities that are necessary to drive recombination?

To directly test the role of transcriptional read-through in gene segment rearrangement, the Krangel laboratory generated *Tcra* knock-in mice with

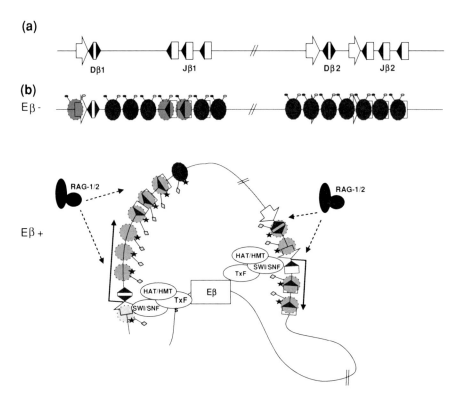

Fig. 2 Model of the epigenetic landscapes that regulate recombinational accessibility across the DJβ1 and DJβ2 cassettes. **a** Schematic of DJβ organization (not drawn to scale). Each Dβ with flanking RSS elements (*blacktriangles*), germ-line promoters (*blockarrows*), and first 3 Jβ segments are shown. **b** In the absence of Eβ (*upperpanel*), both DJβ cassettes are organized into dense nucleosomal arrays (*shadedcircles*) that bear H3K9me and H3K27me3 marks (*black* and *grey* circular *tags*, respectively). Nucleosome abundance at each location is indicated by shading. When Eβ is present (*lowerpanel*) it folds back to help recruit multiple factors including histone acetyltransferases (HAT) and methyltransferases (HMT), SWI/SNF, and multiple transcription factors (T×F) to the Dβ1 and 3'Dβ2 promoters. Eβ induces replacement of H3K9me with H3K9ac (*diamonds*) and leads to promoter activation, which in turn triggers replacement of H3K27me3 with H3K4me3 marks (*stars*), a dramatic reduction in nucleosome density, transcriptional read-through (*blackarrows*), and accessibility of Dβ and Jβ RSSs to RAG-1/2

concatenated arrays of transcriptional terminator elements inserted upstream of Jα61 or downstream of Jα56 (Abarrategui and Krangel 2006). Blockade of transcriptional elongation suppressed *Tcra* recombination and reduced histone acetylation in the untranscribed Jα segments located downstream of the introduced terminator. Impaired recombination has since been directly correlated with reduced RAG-1 binding to Jα gene segments that lie downstream of the terminator (Ji et al. 2010a). These studies established transcriptional read-through as a critical component of the mechanisms that render downstream Jα segments accessible to recombinase.

Genetic and Epigenetic Regulation of *Tcrb* Gene Assembly

In seeming contrast however, rearrangement of *Tcrb* transgenic reporters is unaffected by inversion of PDβ1, a mutation that essentially abolishes transcription through the Dβ1 and Jβ1 segments (Sikes et al. 2002). Given that PDβ1 is essential for DβJβ recombination, the unexpected insensitivity of recombination to PDβ1 inversion suggests that localized Dβ accessibility is not directly dependent on transcriptional read-through. Rather, other promoter-directed activities, such as transcriptional initiation, must underlie PDβ ACE function.

Promoters assemble distinct and complex networks of regulatory factors that allow for cell type- and developmental stage-specific control of neighboring gene expression. However, replacement of PDβ1 with heterologous promoters readily restores DJβ1 recombination in chromosomal *Tcrb* miniloci (Sikes et al. 1999), suggesting that ACE function is not a property unique to the PDβ1 promoter. If recombinational accessibility at DβJβ segments requires a general feature of promoter activation, but not transcriptional elongation, it remains likely that chromatin changes associated with assembly of transcription complexes at PDβ drive RAG access at the nearby Dβ RSS. Indeed, when PDβ1 is moved from its native location (5$'$ of Dβ1) to sites between Dβ1 and Jβ1.1 or 3$'$ of Jβ1.2, the efficiency of D-to-J recombination is severely compromised (Sikes et al. 2002). Thus, PDβ1 acts as a highly localized ACE.

In what may be a related observation, Dβ2-to-Jβ recombination is inefficient relative to Dβ1-to-Jβ joining in DN thymocytes (Born et al. 1985; Haars et al. 1986; Uematsu et al. 1988). Given the highly localized ACE function of Dβ promoter(s), the initial use of a more distal promoter, situated between Dβ2 and the Jβ2 cluster (3$'$PDβ2) (McMillan and Sikes 2008), may provide a satisfying explanation for reduced Dβ2Jβ recombination in ETP/DN2 cells. The skewing of initial rearrangements to the Dβ1Jβ1 cluster is potentially significant for sculpting TCRβ repertoires because delayed formation of DJβ2 joints offers each *Tcrb* allele a second opportunity to rescue out-of-frame Vβ-to-DJβ1 joints by forming an in-frame V(D)Jβ2 rearrangement.

What may be at the heart of localized PDβ1 ACE function? Assembly of a functional transcription complex at promoters involves the recruitment of numerous factors that covalently modify histones (e.g., H3K4me3) and that remodel DNA:histone associations (Li et al. 2007). The C-terminal region of RAG-2 contains a plant homeodomain (PHD) finger that facilitates RAG binding to H3K4me3 (Liu et al. 2007; Matthews et al. 2007), which in turn stimulates the endonuclease function of bound RAG complexes (Shimazaki et al. 2009). Mutations in RAG-2's PHD finger severely compromise recombination and lead to SCID or Omenn Syndromes (Couedel et al. 2010), immunodeficiencies characterized by the failure of lymphocytes to properly recombine antigen receptor genes, suggesting a central role for H3K4me3 in recombinational control. Given that H3K4me3 marks transcriptionally active genes, perhaps PDβ focuses RAG proteins at the Dβ and Jβ RSSs by driving H3K4me3 deposition at these gene segments. Consistent with this model, RAG binding (Ji et al. 2010a) and H3K4me3 levels (Abarrategui and Krangel 2006) are reduced at Jα segments immediately downstream of a heterologous transcriptional terminator.

Despite such findings, it remains unclear whether H3K4me3, in and of itself, is sufficient to render Dβ segments accessible to recombinase.

Promoters and enhancers are particularly potent chromatin reorganization centers that offset local nucleosome packaging, rendering these elements hypersensitive to DNAse I digestion. Both Eβ and PDβ1 overlap DNAse I hypersensitivity sites mapped within *Tcrb* (Chattopadhyay et al. 1998), and PDβ1 disrupts nucleosome organization over a several hundred bp region spanning Dβ1 (Kondilis-Mangum et al. 2010). Given the inhibitory effect of nucleosomes on the ability of RAG proteins to bind RSSs in vitro, promoter-mediated depletion of nucleosomes may be central to recombinase accessibility at Dβ1. The Dβ2 segment, on the other hand, is several kb from the nearest DNAse I hypersensitivity site, and appears to be fully contained within a single nucleosome in DN thymocytes (Kondilis-Mangum et al. 2010). As such, the higher nucleosome density associated with Dβ2 RSSs may further impair RAG binding and contribute to the relative inefficiency of Dβ2 recombination.

Nucleosomes are displaced from active promoters by chromatin remodeling complexes, such as SWI/SNF, which bend promoter DNA and facilitate recruitment of the TFIID complex (Saha et al. 2006). It follows that an essential aspect of promoter ACE function may be its capacity to displace histones from neighboring RSSs. Consistent with this model, tethering of SWI/SNF adjacent to Dβ1 is sufficient to drive DβJβ recombination in the absence of PDβ1 and its collection of cognate transcription factors (Osipovich et al. 2007). The requirement for promoter-mediated recruitment of SWI/SNF to displace Dβ1 nucleosomes may explain two aspects of PDβ1 ACE function: its apparent transcriptional independence and its position-dependence (Sikes et al. 2002). As chromatin remodeling activity is focused further away from the Dβ1 23-RSS, one anticipates that RAG access to the RSS would diminish.

Both D-J-Cβ cassettes are: (1) transcriptionally activated in ETPs (McMillan and Sikes 2008) (2) equivalently enriched for H3/H4ac and H3K4me3 (Ji et al. 2010a, b), and (3) indistinguishably decorated with RAG-1/2 in DN thymocytes (Ji et al. 2010b). Yet as mentioned above, there is a significant difference in the recombination efficiency of these clusters, with DJβ1 joints accumulating much more rapidly than DJβ2 joints (Born et al. 1985; Haars et al. 1986; Uematsu et al. 1988). The disparity between RAG binding and the efficiency of individual DJβ rearrangements could reflect the semi-quantitative nature and limited resolution of RAG-ChIP assays, one would expect the binding of RAG-1 to be lower at Dβ2 than at Dβ1 if accessibility is regulated at the level of RAG recruitment. Consequently, the apparent equivalence of RAG-1 binding at Dβ1 and Dβ2 might be due to difficulties in resolving RAG bound at the Dβ RSSs from RAG bound at the proximal Jβ RSSs positioned only a few hundred bp downstream. Indeed, RAG-1 and RAG-2 binding levels at either Jβ1.1 or Jβ2.1 are equal to or greater than those measured at the corresponding D elements (Ji et al. 2010b). As technologies advance, more refined experiments will be necessary to tease apart RAG occupancy and recombinational competency among the DJβ clusters, findings that

Genetic and Epigenetic Regulation of *Tcrb* Gene Assembly

would be broadly applicable to understanding recombination efficiencies at all antigen receptor loci.

4.3 Promoter-Mediated Accessibility of Vβ-to-DβJβ Recombination

Vβ gene segments are recombinationally inert in ETP and DN2 cells that execute Dβ-to-Jβ joining, and only begin to rearrange at the DN3 stage of thymocyte development (Livak 2004). Compared with DJβ assembly, Vβ-to-DJβ recombination in DN3 cells requires the opening of chromatin over a more extensive array of gene segments, spanning over 600 kb. As noted earlier, deletion of either Eβ or PDβ1 blocks both Dβ1-to-Jβ and Vβ-to-(Dβ1)Jβ recombination (Bories et al. 1996; Bouvier et al. 1996; Whitehurst et al. 1999). However, neither of these *cis* elements directly controls Vβ accessibility or recombination (Mathieu et al. 2000). Rather, their role during Vβ rearrangement is likely restricted to maintaining recombinase accessibility at rearranged DJβ elements.

If chromatin accessibility over the large Vβ cluster is independent of PDβ and Eβ, what is controlling this critical aspect of *Tcrb* gene assembly? The coordinated timing of Vβ expression and rearrangement suggests that, like their Dβ counterparts, Vβ promoters may control accessibility at associated gene segments via chromatin changes that accompany transcriptional activation. Consistent with this model, deletion of the Vβ14 promoter specifically inhibits rearrangements involving Vβ14, a defect that can be rescued with a heterologous promoter (PSV40) (Ryu et al. 2004). Moreover, insertion of the *Tcra* enhancer, Eα, upstream of Vβ15 specifically increases the frequency of Vβ15 recombination in DN3 cells (Jackson et al. 2005). These data suggest that Vβ recombinational accessibility is regionally controlled by individual Vβ promoters. Whether these promoters act independently or in concert with unidentified *Tcrb* enhancers remains to be resolved.

5 Regulating Locus Assembly Beyond 12/23

Tcrb is organized such that 12-RSSs are found 5′ of each Dβ and Jβ segment, and 23-RSSs are positioned 3′ of the Vβ and Dβ segments (Fig. 1). Consequently, Vβ-to-Jβ, Vβ-to-Dβ, and Dβ-to-Dβ rearrangements would all be in accord with 12/23 restrictions. Nonetheless, these alternative rearrangements are observed only rarely in developing thymocytes, suggesting that Dβ inclusion in *Tcrb* rearrangements is enforced at a level beyond 12/23 (Sleckman et al. 2000). Clearly, recombinational accessibility is intimately associated with a gene segment's epigenetic organization. However, in addition to the ACE activities of PDβ1 and Eβ, multiple studies have suggested critical roles for Dβ RSSs as genetic determinants of recombination efficiency, regulating "Beyond 12/23" (B12/23) restrictions

(Tillman et al. 2004). For example, inappropriate Vβ-to-Jβ assembly can be induced when the native RSS flanking either of these gene segments is replaced with a Dβ RSS (e.g., replacing the Vβ 23-RSS with the Dβ1 23-RSS) (Bassing et al. 2000; Wu et al. 2003). Additionally, transfected RSS-containing substrates fully recapitulate B12/23 assembly in non-lymphoid cells and cell-free extracts containing purified RAG proteins, confirming that Dβ RSSs are encoded with the capacity to direct ordered *Tcrb* assembly (Drejer-Teel et al. 2007; Jung et al. 2003).

How does the Dβ RSS enforce B12/23 restriction? Biochemical evidence suggests that RAG proteins first bind at one RSS, and then capture a second, compatible RSS into a synaptic complex (Jones and Gellert 2002; Mundy et al. 2002). The RAG proteins then initiate recombination by inducing ss DNA breaks (nicks) that are resolved through a transesterification reaction into closed hairpin coding ends and blunt signal ends. An analysis of RSS nicks on Igk alleles suggests a "12-RSS first" model in which RAG proteins initially bind an accessible 12-RSS, and then capture a compatible 23-RSS (Curry et al. 2005), a model that would require RAGs to first bind Jβ 12-RSSs and capture a Dβ 23-RSS. How the "12-RSS first" model would accommodate the apparent dominance of the Dβ 23-RSS in regulating B12/23 is unclear. However, two independent studies have recently suggested an alternative model for DβJβ recombinational control. Oligonucleotide capture in DN thymocytes found that *Tcrb* RSS nicks are predominantly found at the 23-RSS flanking Dβ1 and Dβ2, and not at the Jβ 12-RSSs, suggesting that RAG first binds the Dβ 23-RSS, and then captures a Jβ 12-RSS (Franchini et al. 2009). At the same time, ChIP analyses of RAG binding indicate that both Dβ and Jβ RSSs are bound by RAG-1 and RAG-2, while upstream and downstream Vβ elements remain unbound (Ji et al. 2010b). These tightly focused "recombination centers" appear to be the primary sites of RAG recruitment prior to *Tcrb* assembly, and suggest that RAG binding is not restricted to the 12-RSS, but occurs on all RSSs within a recombination center.

While facilitating D-to-J recombination, the Dβ RSSs appear to possess intrinsic targeting activities that further restrict premature Vβ-to-Dβ rearrangement (Franchini et al. 2009). Given the extreme proximity of the Dβ 12- and 23-RSSs (separated by only 12 nucleotides of Dβ coding sequence), preferential binding of RAG-1/2 to the downstream Dβ1 23-RSS may impede access to the upstream Dβ1 12-RSS, thereby restricting Vβ assembly until the Dβ 23-RSS is removed by D-to-J recombination. In what may be a related finding, the AP-1 transcription factor, c-Fos, binds to a site in the Dβ1 23-RSS, and may also hamper RAG cleavage at the Dβ1 12-RSS until after Dβ-to-Jβ joining (Wang et al. 2008). Moreover, biochemical evidence indicates that c-Fos interacts directly with RAG-2, and potentially with RAG-1, perhaps enhancing RAG deposition at the Dβ1 23-RSS, while simultaneously impeding RAG binding at the Dβ 12-RSS. However, it remains unclear whether c-Fos might hinder RAG access to the Dβ1 12-RSS by allosteric or competitive inhibition.

Repression of 5'PDβ2 in germline DJβ2 clusters is mediated by Upstream Stimulatory Factor 1 (USF-1), which binds to a conserved E-box positioned within the 5'Dβ2 12-RSS spacer (McMillan, et al., submitted). Although USF-1 could potentially restrict inappropriate Dβ1-to-Dβ2 recombination via allosteric mechanisms similar to those discussed for c-Fos, its ability to prevent germ-line transcription of Dβ2 may be more significant in explaining the nucleosomal organization and reduced efficiency of DJβ2 assembly (Born et al. 1985; Haars et al. 1986; Kondilis-Mangum et al. 2010). Likewise, evidence suggests that when the Dβ1 23-RSS is removed from its normal chromosomal context near PDβ1, it can no longer impose B12/23 restriction on gene segment assembly, regardless of c-Fos binding (Wu et al. 2003; Yang-Iott et al. 2010). Taken together, these separate findings suggest that Dβ RSSs, particularly the 23-RSSs that mediate DβJβ joining, actively coordinate *Tcrb* recombination efficiencies at a level beyond 12/23, but within the context of a promoter-driven epigenetic landscape.

6 Long-range Changes in Locus Conformation

During antigen receptor gene assembly, two compatible RSSs must join together with the RAG proteins in a synaptic cleavage complex. Whereas Dβ and Jβ elements are closely spaced in the $3'$ portion of *Tcrb*, formation of synaptic complexes between rearranged DJβ elements and upstream Vβs require spanning over 600 kb of linear DNA sequence. Like nonadjacent links in a chain, distant RSS elements can only interact with one another if the chromatin fiber to which they belong is bent or looped to bring them into juxtaposition. Indeed, chromatin fibers are extremely flexible and in continuous motion, necessary properties given that 2 m of eukaryotic DNA must be compacted to fit within a roughly 10 μm nucleus. Euchromatic wrapping of the DNA around histone cores and even further helical stacking of these nucleosomes in the 30 nm heterochromatin fiber only partially accounts for the compaction necessary to accommodate nuclear constraints. Significantly higher levels of compaction are achieved as chromatin fibers undergo repeated dynamic reorganization from relatively localized shifts between euchromatin and heterochromatin to global reorganization of chromosomal arms in dividing cells.

Higher order chromatin structures can be predicted if the chromatin fiber is treated as a polymer of repeating segments that fold via a random walk model. This approach has suggested a variety of potential structures, the most basic of which include worm-like, freely jointed, and self-avoiding chains (Jhunjhunwala et al. 2009). Worm-like chains are continuously flexible, folding upon themselves like strings. In the latter two models inflexible chromatin segments are connected by hinge regions very much like the links of a bicycle chain, which can either fold upon themselves (freely jointed) or fold to avoid intersecting segments (self-avoiding). Although each model offers the potential for dramatic chromatin compaction, none of these free random walk models can account for geometries observed for mitotic chromosomes. More confined models propose that the chromatin fiber is tethered in 3–5 Mb intervals to a flexible backbone (random walk/giant loop model) or in smaller 1–2 Mb intervals to central linker molecules, forming a chain of rosettes (multiloop subcompartment model) (Jhunjhunwala et al. 2009). Indeed, chromatin rosettes have long been observed (Okada and Comings 1979; Paulson and Laemmli 1977), and have been proposed to underlie V segment juxtaposition with downstream DJ segments in both the Igh and *Tcrb* loci (Bonnet et al. 2009; Jhunjhunwala et al. 2008).

Recent high resolution FISH mapping of Igh during early B cell development suggests that chromatin does not always fold with the same uniform periodicity predicted by random walk/giant loop or multiloop subcompartment models (Jhunjhunwala et al. 2008). Prior to lineage commitment, pre-pro-B cells appear to organize Igh into 1 Mb rosettes that are consistent with predictions of the multi-loop subcompartment model. However, upon transition to the pro-B cell stage and the onset of V(D)J recombination, the VH region undergoes a dramatic compac-tion that juxtaposes V segments with the downstream DHJH segments. Based on these findings, the authors conclude that chromatin fibers likely fold into loops and multiloop rosettes of varying size and spacing, and that this folding is extremely dynamic (Jhunjhunwala et al. 2008).

The long-range folding observed for VH gene segments is unlikely to be a unique feature of the *Igh* locus or even of antigen receptor loci in general. Similar long distance intra- and inter-chromosomal interactions are required at multiple points in the life of eukaroytic cells to drive the processes of gene expression, sister chromatid cohesion, and meiotic crossing over, to name a few (Miele and Dekker 2008; Wallace and Felsenfeld 2007). What types of protein factors might then be responsible for anchoring chromatin loops? Chromatid cohesion depends on the recruitment of a ring-like complex of Cohesin proteins that encircle target DNA (Dorsett 2010). Over the last 15 years, Cohesin has also been linked to stem cell maintenance, developmental regulation, and proliferation via interactions with two distinct DNA binding proteins, the CCCTC-binding factor, CTCF, and the transcriptional coactivator, Mediator (Kagey et al. 2010; Parelho et al. 2008; Wendt et al. 2008). Both modes of Cohesin action may be important in the chromosomal partitioning that occurs during antigen receptor gene assembly.

Cohesin–CTCF interactions were first uncovered in studies of CTCF's role as an insulator-binding protein (Wendt et al. 2008). Insulators are positioned throughout the genome where they segregate chromosomal domains, blocking the spread of either repressive or open chromatin and/or preventing promoter–enhancer interactions (Bell et al. 1999; Wallace and Felsenfeld 2007). CTCF has recently been shown to bind a variety of sites throughout the Ig heavy and light chain loci (Degner et al. 2009). With regard to function, a series of CTCF sites upstream of DH segments are predicted to act as insulators that block association of the VH elements with Eμ enhancers (Featherstone et al. 2010). Feeney and colleagues have shown that Cohesin co-localizes at these CTCF sites within Igh (Degner et al. 2009). Complexes of CTCF and Cohesin sprinkled across the 2.5 kb VH domain could well account for Igh loops that form in pro-B cells, compacting the locus into a rosette structure, which brings VH segments into close proximity with the DJH segments, thereby facilitating the formation of synaptic cleavage complexes (Sayegh et al. 2005). While CTCF sites have yet to be identified at

Tcrb, their involvement in Ig gene assembly, the need to insulate Vβ chromatin during DJβ recombination, and the need to bring Vβ RSSs into proximity with the DJβ RSSs make the investigation of CTCF at *Tcrb* a high priority.

All antigen receptor loci, including *Tcrb*, undergo a dramatic contraction coincident with long-range recombination (V-to-DJ or V-to-J). Chromosome conformation capture (3C) and 3D-FISH studies demonstrate that *Tcrb* folds into contractive loops in DN thymocytes, bringing Vβ elements into spatial proximity with DJβ and Eβ elements (Skok et al. 2007). Locus contraction on unrearranged alleles is reversed in DP cells where V-DJβ recombination is prohibited by allelic exclusion. Looping that allows *Tcrb* to form an Igh-like rosette structure would overcome obvious spatial barriers to Vβ:DJβ synapsis. The Krangel laboratory has shown that the Eβ-proximal Vβ31 segment (only 3 kb downstream of Eβ) as well as Vβ segments situated immediately upstream of a functionally rearranged V(D)Jβ exon retain chromatin accessibility in DP cells, though they do not rearrange efficiently at this developmental stage. In contrast, more distal Vβ segments on rearranged alleles are condensed into facultative heterochromatin (Jackson and Krangel 2005). These findings likely reflect the developmentally controlled unfolding of *Tcrb* upon DN to DP transition, which could reinforce allelic exclusion by a combination of several mechanisms, including: (1) releasing upstream Vβ's from the Eβ activation domain, (2) forming a gradient of heterochromatic condensation upstream of the VDJ exon, and (3) spatially segregating the upstream Vβ segments to inhibit synapsis. The precise mechanisms that govern locus contraction and relaxation in developing thymocytes remain an important area for future studies.

In this regard, Cohesin is an attractive candidate for the regulation of *Tcrb* contraction. As stated above, Cohesin interacts with CTCF, as well as the transcriptional coactivator, Mediator. Recent studies suggest that Mediator:Cohesin complexes lacking CTCF facilitate the formation of DNA loops between promoters and enhancers that drive transcriptional activation (Kagey et al. 2010). Indeed, 3C studies show that Eβ loops to directly contact PDβ1 and PDβ2 regions in DN thymocytes (Fig. 2) (Oestreich et al. 2006). If Mediator and Cohesin are found to direct loop formation between Eβ and Dβ segments, they may play similar roles in contracting the Vβ portion of *Tcrb*. It remains of great interest to determine whether Vβ contraction occurs in Eβ-deficient DN thymocytes. Although Vβ transcription appears to be independent of Eβ (Mathieu et al. 2000), a block in *Tcrb* contraction following enhancer deletion would suggest that Eβ does contribute to Vβ recombination, perhaps by recruiting factors like Cohesin and Mediator or even CTCF to drive locus contraction.

7 *Tcrb* Allelic Exclusion

The enormous breadth of our antigen receptor repertoire is generated from large numbers of lymphocytes, each expressing receptors bearing a signature antigen specificity. A potentially harmful byproduct of random receptor generation by

V(D)J recombination is the significant possibility of creating lymphocytes that assemble an autoreactive Ig or TCR. Indeed, multi-layered cellular mechanisms, collectively called tolerance, destroy or render inactive such autoreactive B and T cell clones. The potential for producing autoreactive clones would be even greater if precursor lymphocytes were allowed to fully assemble both alleles of each antigen receptor gene. However, the vast majority of mature lymphocytes exhibit mono-allelic expression of rearranged Ig and *Tcr* genes (Brady et al. 2010). This phenomenon, called allelic exclusion, is analogous to other mono-allelic regulatory processes that are critical for vertebrate development, including X chromosome inactivation and imprinting, which are mediated by epigenetic and transcriptional repression (Zakharova et al. 2009).

Mature B and T cells generally possess a single in-frame VDJ rearrangement at Igh, Igk, or *Tcrb* loci (Brady et al. 2010; Wucherpfennig et al. 2007), suggesting that allelic exclusion prevents assembly of a second functional gene. Indeed, expression of a functionally rearranged *Tcrb* transgene in thymocytes blocks $V\beta$-(D)$J\beta$ rearrangements on both endogenous *Tcrb* alleles (Bluthmann et al. 1988; Uematsu et al. 1988). Assembly of an in-frame V(D)Jβ joint leads to cell surface expression of TCRβ protein as part of the pre-TCR complex, and analyses of *Tcrb* recombination and thymocyte development in pTα-deficient mice demonstrate that pre-TCR signaling is an essential trigger for allelic exclusion (Aifantis et al. 1998).

Although *Tcrb* transgenes effectively block endogenous $V\beta$ recombination, they do not prevent $D\beta$-to-$J\beta$ rearrangement, arguing that allelic exclusion is specifically imposed at the level of $V\beta$ accessibility (Uematsu et al. 1988). However, *Tcrb* is contracted in DN cells, removing the spatial constraints that prevent juxtaposition of upstream $V\beta$ segments with the $DJ\beta$ joints (Skok et al. 2007). Additionally, at least the $V\beta13$-2 segment (formerly $V\beta8.2$) is bi-allelically transcribed in DN2/3 cells prior to $V\beta$ recombination (Jia et al. 2007). If $V\beta$ RSS elements are transcribed and juxtaposed with $DJ\beta$ joints in DN cells, how might *Tcrb* assembly and expression be limited to a single allele? When the Busslinger and Krangel laboratories examined nuclear localization of *Tcrb* domains in DN cells, they found that the distal $V\beta$ region tended to associate with pericentromeric heterochromatin, a profoundly repressive environment (Schlimgen et al. 2008; Skok et al. 2007). Whereas the Busslinger group found this association with heterochromatin to be monoallelic (Skok et al. 2007), the Krangel group found it to be more frequent and stochastic (Schlimgen et al. 2008). They suggest that such stochastic associations may provide a repressive environment necessary to limit bi-allelic $V\beta$ recombination. Consistent with this model, insertion of Eα into the upstream $V\beta$ array reduced the frequency of association with pericentromeric heterochromatin and impaired allelic exclusion (Schlimgen et al. 2008). Together, these studies suggest that the enriched heterochromatic association of $V\beta$ gene segments in DN cells may play a crucial first step in imposing *Tcrb* allelic exclusion by limiting $V\beta$ recombination on the second (heterochromatic) allele.

Genetic and Epigenetic Regulation of *Tcrb* Gene Assembly

Additional factors beyond Vβ proximity must also contribute to *Tcrb* allelic exclusion. Indeed, Vβ31 is positioned immediately 3' of Eβ, which obviates the need for locus contraction in its juxtaposition with DJβ elements. Despite this proximity, allelic exclusion is efficiently imposed on Vβ31 in DN/DP thymocytes. A partial explanation for this contraction-independent regulation may be that Vβ RSS elements are relatively poor substrates for RAG-mediated cleavage in vitro (Jung et al. 2003), and are restricted in vivo to only two potential synaptic cleavage partners (DJβ1 or DJβ2). Indeed, allelic exclusion of Vβ31 can be broken when its RSS is replaced with much more efficient RSSs that flank Dβ1 (Wu et al. 2003; Yang-Iott et al. 2010). In sum, the current evidence supports a model for *Tcrb* allelic exclusion in which both the genetic determinants and epigenetic organization of Vβ elements in DN cells stack the odds against simultaneous bi-allelic Vβ recombination.

If initial Vβ-to-DJβ assembly is stochastic, and reflects one allele's chance reorganization into a recombinationally accessible state before the second allele, how does pre-TCR signaling block additional Vβ rearrangements before the RAG proteins are inactivated and removed? The bHLH protein E47, a splice variant of the E2A protein, binds E box elements at multiple Vβ segments, the DJβ cassettes, and Eβ (Agata et al. 2007). In keeping with this distribution, E47 is critical for 5'PDβ2 activity and Dβ-to-Jβ recombination (Agata et al. 2007; McMillan and Sikes 2009). During β selection, the process by which feedback inhibition leads to allelic exclusion and developmental progression, E47 expression is down-regulated while expression of the antagonistic Id3 protein is up-regulated. The consequence of these related events is that E47 occupancy across *Tcrb* declines during β selection. The Murre laboratory has shown that feedback inhibition through the pre-TCR reduces E47 binding and chromatin accessibility at Vβ promoters. Moreover, enforced expression of E47 specifically increases the frequency of Vβ-to-(D)Jβ but not Dβ-to-Jβ rearrangement in DN cells, and leads to the accumulation of Vβ signal joints in both DN and DP cells (Agata et al. 2007). These data suggest that E47 may not only be critical for Vβ recombinational accessibility, but that its loss during β selection is an essential component of allelic exclusion.

Once imposed, allelic exclusion at *Tcrb* must be maintained in DP cells that are actively rearranging *Tcra*. Whereas *Tcrb* loci are essentially fully contracted in DN cells, unrearranged alleles are decontracted in DP cells, preventing the juxtaposition of DJβ segments with upstream Vβ segments (Skok et al. 2007). When Eα was inserted upstream of Vβ15, its presence led to enhanced Vβ usage in DN cells. In DP cells, the Eα insertion dramatically enhanced Vβ15 chromatin accessibility and germ-line transcription; however, it failed to break allelic exclusion (Jackson et al. 2005). Analyses of RAG binding and decontraction of loci that harbor the Eα knock-in have not been reported. However, decontraction of *Tcrb* in DP cells provides a physical barrier against DJβ synapsis with upstream Vβ elements, perhaps providing an explanation for the control of this Vβ cluster "beyond accessibility". Consistent with this possibility, repositioning of Vβ14 to a site only 6.8 kb upstream of Dβ1 partially subverts allelic exclusion, suggesting that distance from the DJβ elements may be critical for suppressing rearrangement of upstream Vβ segments (Sieh and Chen 2001). Thus, the current evidence suggests that stochastic association with pericentromeric heterochromatin limits *Tcrb* recombination to the other allele. Successful recombination and pre-TCR expression leads to feedback inhibition that triggers chromosomal relaxation and reverses V-DJ juxtaposition on the unrearranged allele in DP cells. Coupled with the sub-optimal recombination efficiency of Vβ RSSs, these chromosomal constraints initiate an allelic exclusion process that is enforced by the loss of E47 and chromatin accessibility. However, numerous questions remain to be resolved. Dissection of the potential mechanisms for mediating *Tcrb* allelic exclusion is likely to remain a research focus for several years to come, as insights drive the need for increasingly robust, refined, and developmentally dynamic assay systems.

8 Conclusions

In the 25 years since Yancopoulos and Alt (1985) first proposed the recombinational accessibility model, a wealth of data has amassed to support their central hypothesis—that germ-line transcription and associated cis-elements guide antigen receptor gene assembly. Technologies unavailable at the time have since shown that antigen receptor loci indeed reorganize their chromatin into accessible and inaccessible configurations during lymphocyte development and alter their three-dimensional structures to bring distal gene segments into spatial proximity (Thomas et al. 2009). We now know that enhancers, promoters, and consequential germ-line transcription are each central to chromatin accessibility and recombination (Abarrategui and Krangel 2009; Osipovich and Oltz 2010). Chromatin accessibility also provides a necessary environment by which genetic differences between RSSs favor the developmental order of gene segment pairing (Curry et al. 2005; Franchini et al. 2009; Ji et al. 2010a). Recent findings have begun to implicate specific transcription factors, chromatin remodelers, and histone modifiers as protein mediators of accessibility (Agata et al. 2007; Osipovich et al. 2007, 2004; Wang et al. 2008). Nevertheless, the mechanisms that stitch these various components into a single program that is informed by and, in turn, informs lymphocyte development have proven elusive.

Emerging studies have also raised questions regarding the nature of antigen receptor chromatin in stem cells that have not yet committed to a lymphocyte lineage. Similar to many genes that encode regulators of growth and development, initial evidence suggests that Igh and *Tcr* genes are primed for activation in embryonic and hematopoietic stem cells. For example, the TEA promoter binds a number of transcription factors in DN thymocytes, even in the absence of Eα (Hernandez-Munain et al. 1999). This promoter is also hypomethylated and associates with H3K4me marks, despite its transcriptional silence (Sikes et al. 2009). In addition, the Vβ14 segment is enriched in hematopoietic stem cells for H3K4me3 and H3K27me3 (Weishaupt et al. 2010). This bivalent marking with activation- and repression-associated modifications appears to be critical for transcriptional priming of developmentally regulated genes prior to their activation and may also prove to be critical for antigen receptor gene assembly. Analysis of the mechanisms that might act well before locus activation as priming events, and how these actions might coordinate with signals specific to a stem cell's micro-environment and developmental state should prove enlightening. In this regard, the Smale laboratory has shown that the pTα enhancer is primed in ES cells, where it is bound by key transcription factors and is protected from CpG methylation (Xu et al. 2009). Importantly, they find that developmental priming is essential for pTα expression later in thymocyte development (Xu et al. 2007), illustrating the significance of epigenetic programs that start long before lineage decisions have been made.

The intimate cross-talk between genetic determinants and their epigenetic environment promises to yield insights that go far beyond antigen receptor assembly. Developmental gene regulation is a central pillar of vertebrate biology, providing the foundation for cellular specialization. A much better understanding of the triggers and programs that direct developmental gene regulation will be critical if we are to realize the enormous potential offered by medical advances like stem cell therapy. As a more mechanistic understanding of the epigenetic programs that guide antigen receptor assembly and instruct lymphocyte development emerges, we are poised to make incisive discoveries about their impacts on health and disease.

Acknowledgments ML Sikes is supported by grants from the National Institute of Allergy and Infectious Diseases (R56AI070848-01A1) and GSK. EM Oltz is supported by grants from NIAID (AI07973208, AI0822402, AI08251702, AI7494502), NCI (CA15669001), and NIEHS (ES019779).

References

Abarrategui I, Krangel MS (2006) Regulation of T cell receptor-alpha gene recombination by transcription. Nat Immunol 7:1109–1115

Abarrategui I, Krangel MS (2009) Germline transcription: a key regulator of accessibility and recombination. Adv Exp Med Biol 650:93–102

Agata Y, Tamaki N, Sakamoto S, Ikawa T, Masuda K, Kawamoto H, Murre C (2007) Regulation of T cell receptor beta gene rearrangements and allelic exclusion by the helix-loop-helix protein, E47. Immunity 27:871–884

Aifantis I, Azogui O, Feinberg J, Saint-Ruf C, Buer J, von Boehmer H (1998) On the role of the pre-T cell receptor in alphabeta versus gammadelta T lineage commitment. Immunity 9:649–655

Bassing CH, Alt FW, Hughes MM, D'Auteuil M, Wehrly TD, Woodman BB, Gartner F, White JM, Davidson L, Sleckman BP (2000) Recombination signal sequences restrict chromosomal V(D)J recombination beyond the 12/23 rule. Nature 405:583–586

Bell AC, West AG, Felsenfeld G (1999) The protein CTCF is required for the enhancer blocking activity of vertebrate insulators. Cell 98:387–396

Bluthmann H, Kisielow P, Uematsu Y, Malissen M, Krimpenfort P, Berns A, von Boehmer H, Steinmetz M (1988) T cell-specific deletion of T cell receptor transgenes allows functional rearrangement of endogenous alpha- and beta-genes. Nature 334:156–159

Bonnet M, Ferrier P, Spicuglia S (2009) Molecular genetics at the T cell receptor beta locus: insights into the regulation of V(D)J recombination. Adv Exp Med Biol 650:116–132

Bories JC, Demengeot J, Davidson L, Alt FW (1996) Gene-targeted deletion and replacement mutations of the T cell receptor beta-chain enhancer: the role of enhancer elements in controlling V(D)J recombination accessibility. Proc Natl Acad Sci U S A 93:7871–7876

Born W, Yague J, Palmer E, Kappler J, Marrack P (1985) Rearrangement of T cell receptor beta-chain genes during T cell development. Proc Natl Acad Sci U S A 82:2925–2929

Bouvier G, Watrin F, Naspetti M, Verthuy C, Naquet P, Ferrier P (1996) Deletion of the mouse T cell receptor beta gene enhancer blocks alphabeta T cell development. Proc Natl Acad Sci U S A 93:7877–7881

Brady BL, Steinel NC, Bassing CH (2010) Antigen receptor allelic exclusion: an update and reappraisal. J Immunol 185:3801–3808

Chattopadhyay S, Whitehurst CE, Schwenk F, Chen J (1998) Biochemical and functional analyses of chromatin changes at the TCR-beta gene locus during CD4$^-$ CD8$^-$ to CD4$^+$ CD8$^+$thymocyte differentiation. J Immunol 160:1256–1267

Cobb RM, Oestreich KJ, Osipovich OA, Oltz EM (2006) Accessibility control of V(D)J recombination. Adv Immunol 91:45–109

Couedel C, Roman C, Jones A, Vezzoni P, Villa A, Cortes P (2010) Analysis of mutations from SCID and Omenn syndrome patients reveals the central role of the Rag2 PHD domain in regulating V(D)J recombination. J Clin Invest 120:1337–1344

Cowell LG, Davila M, Yang K, Kepler TB, Kelsoe G (2003) Prospective estimation of recombination signal efficiency and identification of functional cryptic signals in the genome by statistical modeling. J Exp Med 197:207–220

Curry JD, Geier JK, Schlissel MS (2005) Single-strand recombination signal sequence nicks in vivo: evidence for a capture model of synapsis. Nat Immunol 6:1272–1279

Degner SC, Wong TP, Jankevicius G, Feeney AJ (2009) Cutting edge: developmental stage-specific recruitment of cohesin to CTCF sites throughout immunoglobulin loci during B lymphocyte development. J Immunol 182:44–48

Dorsett D (2010) Gene regulation: the cohesin ring connects developmental highways. Curr Biol 20:R886–R888

Doty RT, Xia D, Nguyen SP, Hathaway TR, Willerford DM (1999) Promoter element for transcription of unrearranged T cell receptor beta-chain gene in pro- T cells. Blood 93:3017–3025

Drejer-Teel AH, Fugmann SD, Schatz DG (2007) The beyond 12/23 restriction is imposed at the nicking and pairing steps of DNA cleavage during V(D)J recombination. Mol Cell Biol 27:6288–6299

Eastman QM, Leu TM, Schatz DG (1996) Initiation of V(D)J recombination in vitro obeying the 12/23 rule. Nature 380:85–88

Featherstone K, Wood AL, Bowen AJ, Corcoran AE (2010) The mouse immunoglobulin heavy chain V-D intergenic sequence contains insulators that may regulate ordered V(D)J recombination. J Biol Chem 285:9327–9338

Feeney AJ (2009) Genetic and epigenetic control of V gene rearrangement frequency. Adv Exp Med Biol 650:73–81

Franchini DM, Benoukraf T, Jaeger S, Ferrier P, Payet-Bornet D (2009) Initiation of V(D)J recombination by Dbeta-associated recombination signal sequences: a critical control point in TCRbeta gene assembly. PLoS ONE 4:e4575

Golding A, Chandler S, Ballestar E, Wolffe AP, Schlissel MS (1999) Nucleosome structure completely inhibits in vitro cleavage by the V(D)J recombinase. Embo J 18:3712–3723

Haars R, Kronenberg M, Gallatin WM, Weissman IL, Owen FL, Hood L (1986) Rearrangement and expression of T cell antigen receptor and gamma genes during thymic development. J Exp Med 164:1–24

Halle JP, Haus-Seuffert P, Woltering C, Stelzer G, Meisterernst M (1997) A conserved tissue-specific structure at a human T cell receptor beta-chain core promoter. Mol Cell Biol 17:4220–4229

Hayday AC, Pennington DJ (2007) Key factors in the organized chaos of early T cell development. Nat Immunol 8:137–144

Hernandez-Munain C, Sleckman BP, Krangel MS (1999) A developmental switch from TCR delta enhancer to TCR alpha enhancer function during thymocyte maturation. Immunity 10:723–733

Hublitz P, Albert M, Peters AH (2009) Mechanisms of transcriptional repression by histone lysine methylation. Int J Dev Biol 53:335–354

Jackson A, Kondilis HD, Khor B, Sleckman BP, Krangel MS (2005) Regulation of T cell receptor beta allelic exclusion at a level beyond accessibility. Nat Immunol 6:189–197

Jackson AM, Krangel MS (2005) Allele-specific regulation of TCR beta variable gene segment chromatin structure. J Immunol 175:5186–5191

Jankovic M, Nussenzweig A, Nussenzweig MC (2007) Antigen receptor diversification and chromosome translocations. Nat Immunol 8:801–808

Jenuwein T, Allis CD (2001) Translating the histone code. Science 293:1074–1080

Jhunjhunwala S, van Zelm MC, Peak MM, Cutchin S, Riblet R, van Dongen JJ, Grosveld FG, Knoch TA, Murre C (2008) The 3D-structure of the immunoglobulin heavy-chain locus: implications for long-range genomic interactions. Cell 133:265–279

Jhunjhunwala S, van Zelm MC, Peak MM, Murre C (2009) Chromatin architecture and the generation of antigen receptor diversity. Cell 138:435–448

Ji Y, Little AJ, Banerjee JK, Hao B, Oltz EM, Krangel MS, Schatz DG (2010a) Promoters, enhancers, and transcription target RAG1 binding during V(D)J recombination. J Exp Med

Ji Y, Resch W, Corbett E, Yamane A, Casellas R, Schatz DG (2010b) The in vivo pattern of binding of RAG1 and RAG2 to antigen receptor loci. Cell 141:419–431

Jia J, Kondo M, Zhuang Y (2007) Germline transcription from T cell receptor Vbeta gene is uncoupled from allelic exclusion. EMBO J 26:2387–2399

Jones JM, Gellert M (2002) Ordered assembly of the V(D)J synaptic complex ensures accurate recombination. Embo J 21:4162–4171

Jung D, Bassing CH, Fugmann SD, Cheng HL, Schatz DG, Alt FW (2003) Extrachromosomal recombination substrates recapitulate beyond 12/23 restricted VDJ recombination in nonlymphoid cells. Immunity 18:65–74

Kagey MH, Newman JJ, Bilodeau S, Zhan Y, Orlando DA, van Berkum NL, Ebmeier CC, Goossens J, Rahl PB, Levine SS, Taatjes DJ, Dekker J, Young RA (2010) Mediator and cohesin connect gene expression and chromatin architecture. Nature 467:430–435

Khorasanizadeh S (2004) The nucleosome: from genomic organization to genomic regulation. Cell 116:259–272

Kondilis-Mangum HD, Cobb RM, Osipovich O, Srivatsan S, Oltz EM, Krangel MS (2010) Transcription-dependent mobilization of nucleosomes at accessible TCR gene segments in vivo. J Immunol 184:6970–6977

Krangel MS (2009) Mechanics of T cell receptor gene rearrangement. Curr Opin Immunol 21:133–139

Kwon J, Morshead KB, Guyon JR, Kingston RE, Oettinger MA (2000) Histone acetylation and hSWI/SNF remodeling act in concert to stimulate V(D)J cleavage of nucleosomal DNA. Mol Cell 6:1037–1048

Lee JS, Smith E, Shilatifard A (2010) The language of histone crosstalk. Cell 142:682–685

Li B, Carey M, Workman JL (2007) The role of chromatin during transcription. Cell 128:707–719

Liu Y, Subrahmanyam R, Chakraborty T, Sen R, Desiderio S (2007) A plant homeodomain in RAG-2 that binds Hypermethylated lysine 4 of histone H3 is necessary for efficient antigen-receptor-gene rearrangement. Immunity 27:561–571

Livak F (2004) In vitro and in vivo studies on the generation of the primary T cell receptor repertoire. Immunol Rev 200:23–35

Mathieu N, Hempel WM, Spicuglia S, Verthuy C, Ferrier P (2000) Chromatin remodeling by the T cell receptor (TCR)-beta gene enhancer during early T cell development: implications for the control of TCR- beta locus recombination. J Exp Med 192:625–636

Matthews AG, Kuo AJ, Ramon-Maiques S, Han S, Champagne KS, Ivanov D, Gallardo M, Carney D, Cheung P, Ciccone DN, Walter KL, Utz PJ, Shi Y, Kutateladze TG, Yang W, Gozani O, Oettinger MA (2007) RAG2 PHD finger couples histone H3 lysine 4 trimethylation with V(D)J recombination. Nature 450:1106–1110

McMillan RE, Sikes ML (2008) Differential activation of dual promoters alters Dbeta2 germline transcription during thymocyte development. J Immunol 180:3218–3228

McMillan RE, Sikes ML (2009) Promoter activity 5′ of Dbeta2 is coordinated by E47, Runx1, and GATA-3. Mol Immunol 46:3009–3017

Miele A, Dekker J (2008) Long-range chromosomal interactions and gene regulation. Mol Biosyst 4:1046–1057

Morshead KB, Ciccone DN, Taverna SD, Allis CD, Oettinger MA (2003) Antigen receptor loci poised for V(D)J rearrangement are broadly associated with BRG1 and flanked by peaks of histone H3 dimethylated at lysine 4. Proc Natl Acad Sci U S A 100:11577–11582

Mundy CL, Patenge N, Matthews AG, Oettinger MA (2002) Assembly of the RAG1/RAG2 synaptic complex. Molecular & Cellular Biology 22:69–77

Murre C (2009) Developmental trajectories in early hematopoiesis. Genes Dev 23:2366–2370

Oestreich KJ, Cobb RM, Pierce S, Chen J, Ferrier P, Oltz EM (2006) Regulation of TCRbeta gene assembly by a promoter/enhancer holocomplex. Immunity 24:381–391

Okada TA, Comings DE (1979) Higher order structure of chromosomes. Chromosoma 72:1–14

Osipovich O, Cobb RM, Oestreich KJ, Pierce S, Ferrier P, Oltz EM (2007) Essential function for SWI-SNF chromatin-remodeling complexes in the promoter-directed assembly of *Tcrb* genes. Nat Immunol 8:809–816

Osipovich O, Milley R, Meade A, Tachibana M, Shinkai Y, Krangel MS, Oltz EM (2004) Targeted inhibition of V(D)J recombination by a histone methyltransferase. Nat Immunol 5:309–316

Osipovich O, Oltz EM (2010) Regulation of antigen receptor gene assembly by genetic-epigenetic crosstalk. Semin Immunol 22:313–322

Parelho V, Hadjur S, Spivakov M, Leleu M, Sauer S, Gregson HC, Jarmuz A, Canzonetta C, Webster Z, Nesterova T, Cobb BS, Yokomori K, Dillon N, Aragon L, Fisher AG, Merkenschlager M (2008) Cohesins functionally associate with CTCF on mammalian chromosome arms. Cell 132:422–433

Paulson JR, Laemmli UK (1977) The structure of histone-depleted metaphase chromosomes. Cell 12:817–828

Porritt HE, Rumfelt LL, Tabrizifard S, Schmitt TM, Zuniga-Pflucker JC, Petrie HT (2004) Heterogeneity among DN1 prothymocytes reveals multiple progenitors with different capacities to generate T cell and non- T cell lineages. Immunity 20:735–745

Ryu CJ, Haines BB, Lee HR, Kang YH, Draganov DD, Lee M, Whitehurst CE, Hong HJ, Chen J (2004) The T cell receptor beta variable gene promoter is required for efficient V beta rearrangement but not allelic exclusion. Mol Cell Biol 24:7015–7023

Saha A, Wittmeyer J, Cairns BR (2006) Chromatin remodelling: the industrial revolution of DNA around histones. Nat Rev Mol Cell Biol 7:437–447

Sayegh CE, Jhunjhunwala S, Riblet R, Murre C (2005) Visualization of looping involving the immunoglobulin heavy-chain locus in developing B cells. Genes Dev 19:322–327

Schatz DG, Baltimore D (1988) Stable expression of immunoglobulin gene V(D)J recombinase activity by gene transfer into 3T3 fibroblasts. Cell 53:107–115

Schatz DG, Spanopoulou E (2005) Biochemistry of V(D)J recombination. Curr Top Microbiol Immunol 290:49–85

Schlimgen RJ, Reddy KL, Singh H, Krangel MS (2008) Initiation of allelic exclusion by stochastic interaction of *Tcrb* alleles with repressive nuclear compartments. Nat Immunol 9:802–809

Shimazaki N, Tsai AG, Lieber MR (2009) H3K4me3 stimulates the V(D)J RAG complex for both nicking and hairpinning in trans in addition to tethering in cis: implications for translocations. Mol Cell 34:535–544

Sieh P, Chen J (2001) Distinct control of the frequency and allelic exclusion of the V beta gene rearrangement at the TCR beta locus. J Immunol 167:2121–2129

Sikes ML, Bradshaw JM, Ivory WT, Lunsford JL, McMillan RE, Morrison CR (2009) A streamlined method for rapid and sensitive chromatin immunoprecipitation. J Immunol Methods 344:58–63

Sikes ML, Gomez RJ, Song J, Oltz EM (1998) A developmental stage-specific promoter directs germline transcription of D beta J beta gene segments in precursor T lymphocytes. J Immunol 161:1399–1405

Sikes ML, Meade A, Tripathi R, Krangel MS, Oltz EM (2002) Regulation of V(D)J recombination: a dominant role for promoter positioning in gene segment accessibility. Proc Natl Acad Sci U S A 99:12309–12314

Sikes ML, Suarez CC, Oltz EM (1999) Regulation of V(D)J recombination by transcriptional promoters. Mol Cell Biol 19:2773–2781

Skok JA, Gisler R, Novatchkova M, Farmer D, de Laat W, Busslinger M (2007) Reversible contraction by looping of the *Tcra* and *Tcrb* loci in rearranging thymocytes. Nat Immunol 8:378–387

Sleckman BP, Bassing CH, Hughes MM, Okada A, D'Auteuil M, Wehrly TD, Woodman BB, Davidson L, Chen J, Alt FW (2000) Mechanisms that direct ordered assembly of T cell receptor beta locus V, D, and J gene segments. Proc Natl Acad Sci U S A 97:7975–7980

Stanhope-Baker P, Hudson KM, Shaffer AL, Constantinescu A, Schlissel MS (1996) Cell type-specific chromatin structure determines the targeting of V(D)J recombinase activity in vitro. Cell 85:887–897

Steen SB, Gomelsky L, Speidel SL, Roth DB (1997) Initiation of V(D)J recombination in vivo: role of recombination signal sequences in formation of single and paired double-strand breaks. Embo J 16:2656–2664

Swanson PC, Desiderio S (1998) V(D)J recombination signal recognition: distinct, overlapping DNA–protein contacts in complexes containing RAG1 with and without RAG2. Immunity 9:115–125

Swanson PC, Kumar S, Raval P (2009) Early steps of V(D)J rearrangement: insights from biochemical studies of RAG-RSS complexes. Adv Exp Med Biol 650:1–15

Thomas LR, Cobb RM, Oltz EM (2009) Dynamic regulation of antigen receptor gene assembly. Adv Exp Med Biol 650:103–115

Tillman RE, Wooley AL, Hughes MM, Khor B, Sleckman BP (2004) Regulation of T cell receptor beta-chain gene assembly by recombination signals: the beyond 12/23 restriction. Immunol Rev 200:36–43

Tripathi R, Jackson A, Krangel MS (2002) A change in the structure of Vβ chromatin associated with TCR β allelic exclusion. Journal of Immunology 168:2316–2324

Tripathi RK, Mathieu N, Spicuglia S, Payet D, Verthuy C, Bouvier G, Depetris D, Mattei MG, Hempe LW, Ferrier P (2000) Definition of a T cell receptor beta gene core enhancer of V(D)J recombination by transgenic mapping. Mol Cell Biol 20:42–53

Uematsu Y, Ryser S, Dembic Z, Borgulya P, Krimpenfort P, Berns A, von Boehmer H, Steinmetz M (1988) In transgenic mice the introduced functional T cell receptor beta gene prevents expression of endogenous beta genes. Cell 52:831–841

Van Ness BG, Weigert M, Coleclough C, Mather EL, Kelley DE, Perry RP (1981) Transcription of the unrearranged mouse C kappa locus: sequence of the initiation region and comparison of activity with a rearranged V kappa–C kappa gene. Cell 27:593–602

Wallace JA, Felsenfeld G (2007) We gather together: insulators and genome organization. Curr Opin Genet Dev 17:400–407

Wang X, Xiao G, Zhang Y, Wen X, Gao X, Okada S, Liu X (2008) Regulation of *Tcrb* recombination ordering by c-Fos-dependent RAG deposition. Nat Immunol 9:794–801

Weishaupt H, Sigvardsson M, Attema JL (2010) Epigenetic chromatin states uniquely define the developmental plasticity of murine hematopoietic stem cells. Blood 115:247–256

Wendt KS, Yoshida K, Itoh T, Bando M, Koch B, Schirghuber E, Tsutsumi S, Nagae G, Ishihara K, Mishiro T, Yahata K, Imamoto F, Aburatani H, Nakao M, Imamoto N, Maeshima K,

Shirahige K, Peters JM (2008) Cohesin mediates transcriptional insulation by CCCTC-binding factor. Nature 451:796–801

Whitehurst CE, Chattopadhyay S, Chen J (1999) Control of V(D)J recombinational accessibility of the D beta 1 gene segment at the TCR beta locus by a germline promoter. Immunity 10:313–322

Wu C, Bassing CH, Jung D, Woodman BB, Foy D, Alt FW (2003) Dramatically increased rearrangement and peripheral representation of Vbeta14 driven by the 3′Dbeta1 recombination signal sequence. Immunity 18:75–85

Wucherpfennig KW, Allen PM, Celada F, Cohen IR, De Boer R, Garcia KC, Goldstein B, Greenspan R, Hafler D, Hodgkin P, Huseby ES, Krakauer DC, Nemazee D, Perelson AS, Pinilla C, Strong RK, Sercarz EE (2007) Polyspecificity of T cell and B cell receptor recognition. Semin Immunol 19:216–224

Xu J, Pope SD, Jazirehi AR, Attema JL, Papathanasiou P, Watts JA, Zaret KS, Weissman IL, Smale ST (2007) Pioneer factor interactions and unmethylated CpG dinucleotides mark silent tissue-specific enhancers in embryonic stem cells. Proc Natl Acad Sci U S A 104: 12377–12382

Xu J, Watts JA, Pope SD, Gadue P, Kamps M, Plath K, Zaret KS, Smale ST (2009) Transcriptional competence and the active marking of tissue-specific enhancers by defined transcription factors in embryonic and induced pluripotent stem cells. Genes Dev 23:2824–2838

Yancopoulos GD, Alt FW (1985) Developmentally controlled and tissue-specific expression of unrearranged VH gene segments. Cell 40:271–281

Yang-Iott KS, Carpenter AC, Rowh MA, Steinel N, Brady BL, Hochedlinger K, Jaenisch R, Bassing CH (2010) TCR beta feedback signals inhibit the coupling of recombinationally accessible V beta 14 segments with DJ beta complexes. J Immunol 184:1369–1378

Zakharova IS, Shevchenko AI, Zakian SM (2009) Monoallelic gene expression in mammals. Chromosoma 118:279–290

T-Cell Identity and Epigenetic Memory

Ellen V. Rothenberg and Jingli A. Zhang

Abstract T-cell development endows cells with a flexible range of effector differentiation options, superimposed on a stable core of lineage-specific gene expression that is maintained while access to alternative hematopoietic lineages is permanently renounced. This combination of features could be explained by environmentally responsive transcription factor mobilization overlaying an epigenetically stabilized base gene expression state. For example, "poising" of promoters could offer preferential access to T-cell genes, while repressive histone modifications and DNA methylation of non-T regulatory genes could be responsible for keeping non-T developmental options closed. Here, we critically review the evidence for the actual deployment of epigenetic marking to support the stable aspects of T-cell identity. Much of epigenetic marking is dynamically maintained or subject to rapid modification by local action of transcription factors. Repressive histone marks are used in gene-specific ways that do not fit a simple, developmental lineage-exclusion hierarchy. We argue that epigenetic analysis may achieve its greatest impact for illuminating regulatory biology when it is used to locate cis-regulatory elements by catching them in the act of mediating regulatory change.

E. V. Rothenberg (✉) · J. A. Zhang
Division of Biology 156-29, California Institute of Technology,
Pasadena, CA 91125, USA
e-mail: evroth@its.caltech.edu

Contents

1	The Problem of T-Cell Identity	118
2	Combinatoriality	120
3	Epigenetics and Hit-and-Run Gene Regulation	121
4	Interplay Between Cellular History and Current Activation State: Examples from Cytokine Gene Control	122
5	Reversibility and Plasticity of Histone Marks	123
6	Combinatorial Transcription Factor Action and Epigenetic Modification	126
7	A Problem of Repression	128
8	Epigenetic Marking Events in T-Lineage Gene Activation from Stem-Cell Precursors	130
9	Epigenetic Repression of Non-T-Cell Genes During T-Lineage Commitment	132
10	Concluding Remarks: Open Prospects	136
References		138

1 The Problem of T-Cell Identity

Cohorts of T-lineage cells develop from hematopoietic precursors throughout fetal and much of postnatal life in mammals. Basic T-lineage properties, including the gene rearrangements leading to expression of a clonally individual T-cell receptor for antigen (TCR), are conferred by differentiation in the thymus (T lineage commitment). However, T-cell development also continues after these properties are established. Not only are mature T cells long-lived cells with extensive proliferative potential, but also they continue to specialize after leaving the thymus. In response to antigen, they select and mobilize any of a variety of gene expression programs for effector responses, and then reinforce these programs for preferential access during stimulation events in the future (effector subset commitment). The distinct effector programs that mature T-cells can deploy in response to antigen challenge (O'Shea and Paul 2010; Zhou et al. 2009; Zhu et al. 2010; Spits and Di Santo 2011) are of great medical significance, since they determine not only what the T cells are likely to do, but also what intercellular interaction molecules they will express to influence the functions of other immune cells. An early split, occurring within the thymus, separates the mostly $CD4^+$ $TCR\alpha\beta^+$ "helper" cells from the mostly $CD8^+$ $TCR\alpha\beta^+$ "cytotoxic" cells. Then among the $CD4^+$ cells, antigen-triggered functional specialization generates distinct cytokine-producing T-cell subsets which have been designated Th1, Th2, Th17, and Treg cells. In each of these major effector programs, the $CD4^+$ T-cell has its differentiation guided by a combination of subset-specific cytokine receptor signaling, subset-specific STAT factor mobilization, and subset-specific "master" transcription factor action [reviewed by (O'Shea and Paul 2010; Zhu et al. 2010; Murphy and Stockinger 2010; Wilson et al. 2009; Zhou et al. 2009)]. Thus, a set of divergent gene networks can be mobilized for antigen-dependent differentiation, each influenced by intrinsic, autocrine, and paracrine environmental effects. There is considerable interest in how stably or reversibly cells commit to any of these subsets, and some of the emerging answers are reviewed briefly below.

There are actually multiple subtypes of T cells besides these major CD4$^+$ TCR$\alpha\beta^+$ "Th" subsets. Some are additional variants of CD4$^+$ effectors, such as Th9, Th22, and Tfh cells, which may diverge from the other effector types through mechanisms that are under intense discussion (Murphy and Stockinger 2010; Spits and Di Santo 2011). In addition, several more divergent T-cell lineages branch off from all major TCR$\alpha\beta$ lineages in the thymus. These include TCR$\gamma\delta$ subsets that not only use different genes to encode their receptors but also have distinctive homing, cytokine expression, and response threshold properties; and in addition, NKT and possibly also CD8$\alpha\alpha$ "innate type" T cells that follow different triggering rules from conventional TCR$\alpha\beta$ CD4$^+$ T cells (Meyer et al. 2010; Park et al. 2010; Kreslavsky et al. 2009; Spits and Di Santo 2011; Das et al. 2010; Gangadharan et al. 2006). The initial determinant spurring choice of one of these developmental pathways is often the particular TCR the cell expresses, and its interaction with particular ligands expressed in the thymic environment. Nevertheless, the result in each case is to assemble within the cell a particular gene regulatory network that will be the framework for all the cells' responses to future stimulation. All these diverse variants of T cells are testimony to the versatility and dynamism of "the T cell program". However, the developmental problem posed by T-cell differentiation is actually broader and even more challenging than the distinctions among these subsets.

Despite the flexibility, the initial programming that enables a precursor to become a T-lineage cell at all creates a major core identity that is not flexible, and is apparently irreversibly set. The activation-dependent specialization differences are superimposed on this background of stability. Not only do rearranged TCR genes stay rearranged (as they must), but the cells also faithfully preserve a stable program of lineage-specific gene expression that has nothing to do with gene rearrangement, maintaining transcription of genes that encode the invariant TCR complex signaling components CD3γ, δ, ε and TCRζ, the lineage-specific kinases Lck, ZAP70, and Itk, and the crucial signaling adaptor molecules Lat, SLP76 (Lcp2), and GADS (Grap2). All these genes are largely or completely T-cell specific in their RNA expression (http://www.immgene.org) (Heng et al. 2008). Thus, they must be maintained by a specific aspect of T-lineage regulatory state that is held constant, even while the cells dynamically alter their transcriptional program choices in response to their environments. The contrast between stable T-lineage identity functions and versatile, multi-option, actively modulated effector functions is much more obvious in the case of T cells than in the case of B cells. Moreover, wherever they migrate and whatever signals they encounter, mature T cells do not regain access to genes associated with some of the alternative developmental programs that were available to their early precursors, such as the myeloid or B-cell programs.[1] Thus, some fundamental

---

[1] An interesting partial exception is the progenitor cell, B cell, and myeloid cell associated transcription factor PU.1, discussed below.

aspects of the T-cell developmental program are not only established earlier, but also established much more robustly, than effector function specialization. How does this work?

In this review, we start with two general models, combinatoriality and hit-and-run regulation, for understanding the T-cell state. The take-home lessons from well-established paradigms of effector gene expression in later T-cell development are reviewed. We then evaluate more critically how transcription factor action and epigenetic modification intersect, using more recent data from genome-wide analyses. Finally, the principles that emerge are applied to consider the current evidence for the regulatory origins of the T-cell gene expression program, and the crucial questions that need to be tackled in the future.

2 Combinatoriality

Stability of some gene expression patterns while others are changing can be explained simply by the use of different transcription factor combinations to control stable and dynamically regulated genes. Gene expression as a rule depends on combinations of factors both in development and in physiological responses to activation signals, and the precise combinations of inputs needed are dictated by the structure of gene-specific cis-regulatory elements. Thus, whether or not a particular factor is rate-limiting for activation of a given target gene generally depends on what other factors are available to collaborate, and a different requirement may apply at other target genes of the same factor. An extreme case is that of the T-cell cytokine gene, *Il2*, which depends on a complex ensemble of differentially activated transcription factors, all of which can make rate-limiting contributions to its expression (Bunting et al. 2006; Rothenberg and Ward 1996; Jain et al. 1995). Like other transiently activated cytokine genes, it requires inputs from the acutely mobilized signal-response transcription factors AP-1, NFAT, and an NF-κB family member such as c-Rel. When *Il2* is induced, transcription terminates almost immediately when cyclosporin A or FK506 is used to interrupt the availability of one of these factors, and in these cases the entire transcription factor ensemble rapidly dissociates from the promoter-proximal enhancer of the gene (Garrity et al. 1994; Chen and Rothenberg 1994; Rao et al. 2001). Thus, if it were the case that all T cells stably expressed certain core transcription factors, then it is theoretically possible that any genes that need to be expressed stably could simply require combinations of these constant factors. Then, in this scenario, genes involved in dynamic subset specialization would be distinguished by their requirement for different combinations of transcription factors which might need to include at least one factor that was not part of the core, but rather dependent on context or environmental signals. Indeed, this is a way that STAT factors are used to regulate T-effector subset differentiation genes, as noted below. Such a combinatorial gene control model is consistent with known features of gene regulation

(Davidson 2006) and does not necessarily require any hierarchy in mechanisms leading to different degrees of expression stability among different genes.

3 Epigenetics and Hit-and-Run Gene Regulation

There is an alternative way that some gene expression "decisions" could be made more permanent than others, however, and over the recent years this alternative has attracted great interest. The constraint that some transcription factors need to be expressed with perfect stability or excluded continuously can be removed if the genome itself can be selectively masked by modifications of chromatin that are passively maintained from generation to generation. If intrathymic differentiation could position specific epigenetic modifications so as to block or favor particular genes for expression in the future, then these components of the T-cell program would be selectively buffered against regulatory change while others could vary freely during immune responses.

Epigenetic mechanisms were first described to explain the hit-and-run action of the factors that establish Hox complex gene expression patterns, which are then sustained for long-term organization of body plans in embryonic development of many kinds of animals (Grimaud et al. 2006; Schwartz and Pirrotta 2007). Epigenetic modifications of chromatin domains are also found associated with the long-term silencing of repeated gene arrays (Garrick et al. 1998), and implicated in position effect variegation of transgenes inserted near heterochromatin (Williams et al. 2008). Several examples in T-cell gene regulation have set additional precedents for the ability of epigenetic regulatory mechanisms to mediate durable changes in gene accessibility, raising or lowering these genes' thresholds for future activation. Silencing of CD4 expression in the $CD8^+$ lineage depends on Runx factor repressive activity within the thymus, but after the cells are mature, repression of CD4 is less sensitive to Runx/silencer interactions (Taniuchi et al. 2002; Telfer et al. 2004). Conversely, while activation of CD4 in mature $CD4^+$ lineage cells depends on a proximal enhancer, once maturation has occurred CD4 expression continues even if this enhancer is deleted (Chong et al. 2010). Effector response genes in memory $CD8^+$ cells become easier to activate than in naïve T cells also, due to epigenetic modifications of the chromatin at these loci that preserve a partially open state even when the genes are not currently induced (Araki et al. 2009, 2008). The idea of a hit-and-run mechanism for cell-type specification would be particularly appealing to explain T-cell differentiation, because the key events in the establishment of a T-cell identity occur during interaction with a transient Notch pathway signal in the thymus, which is then discontinued and becomes dispensable after commitment (Petrie and Zuniga-Pflucker 2007; Rothenberg et al. 2008).

Well-studied examples of epigenetic changes occur during commitment of a mature, antigen-activated $CD4^+$ T cell to one or another effector subtype (Cuddapah et al. 2010; Murphy and Stockinger 2010; Amsen et al. 2009; Wei et al.

2009; Wilson et al. 2009; Nakayama and Yamashita 2009; Ansel et al. 2006; Reiner et al. 2003), as summarized below. During CD4$^+$ T cell effector polarization, genes of the favored pathways become easier to activate and those of the disfavored pathway become harder to activate than in the naïve cells. In these cases, clear shifts are seen in the patterns of CpG DNA methylation and in chromatin compaction restricting DNase sensitivity around subset-specific cytokine genes. In addition, covalent modification of histone proteins in nucleosomes appear to mediate many of the effects of transcription factors on chromatin, enabling prior transcription factor activity to influence future nucleosome conformation, DNA methylation, and ultimately gene expression.

Most histone marks are thought to favor or reinforce the regulatory effects (activation, repression, etc.) in the course of which they were first deposited (Kouzarides 2007). Thus they are thought to stabilize the induced gene expression pattern in a kind of positive feedback, helping to make differentiation irreversible. "Activating" marks like histone H3 K(9,14) acetylation or H3 K4 trimethylation, "accessibility" marks like H3 K4 mono- or di-methylation, and "repressive" marks like H3K27 trimethylation or H3K9 trimethylation can then affect the tightness of nucleosome packing and the ease of access of transcription factors or RNA polymerase II to their target sequences in the DNA. "Repressive" histone marks also recruit DNA methyltransferases which strengthen repression by methylating local CpG residues. Not only is this associated with recruitment of chromatin condensation proteins, but also it can directly block future recognition of target sites by stimulatory transcription factors (Polansky et al. 2010; Maier et al. 2003). At the same time, combinations of "activating" and "repressive" histone marks can occur at the same loci ("bivalent" marks), and these are strongly associated with instability of expression, perhaps increasing the sensitivity of the gene to future activating or repressive signals (Cui et al. 2009; Bernstein et al. 2006).

4 Interplay Between Cellular History and Current Activation State: Examples from Cytokine Gene Control

Cytokine gene regulation in mature T cells has yielded strong case studies for how epigenetic modifications can mediate the interplay between prior and current transcription factor activity. This extensive literature has been authoritatively reviewed (Wilson et al. 2009; Cuddapah et al. 2010; Murphy and Stockinger 2010; Balasubramani et al. 2010; Zhu et al. 2010; Amsen et al. 2009; Ansel et al. 2006) and is only briefly summarized here. The main take-home lessons have come from the Th2 cytokine gene cluster, including *Il4*, *Il13*, and *Il5*, and from the extensive array of elements that regulate the signature Th1 cytokine gene, *Ifng*. Both the Th2 and the Th1 cytokine loci are activated by TCR signaling, but in mutually exclusive patterns in polarized effector cells. In contrast, naïve T cells respond to TCR activation by activating other loci preferentially, such as the *Il2* locus; they can only weakly and slowly induce expression of either the Th1 or Th2 signature cytokine loci. Yet the transcription factors activated by TCR signaling that directly induce all these genes, including the Ca^{2+}/calcineurin-dependent factor NFAT, are apparently the same in all cases. Thus, the difference in inducibility among Th1 cytokine, Th2 cytokine, and *Il2* loci appears to reside in the prior accessibility marking that these loci undergo during differentiation from naïve to antigen-activated polarized cells.

The IL-4 mobilized transcription factor Stat6 plays a major role in establishing the permissiveness of the Th2 cytokine gene complex for activation (Lee and Rao 2004; Wei et al. 2010), which is then sustained by binding by the product of another Stat6 target gene, the Stat6-upregulated transcription factor GATA-3 (Onodera et al. 2010; Ouyang et al. 2000; Yamashita et al. 2004). As cells differentiate into Th2 effectors, new, subset-specific DNase hypersensitive sites appear that are opened and maintained by Stat6 and GATA-3 binding. Importantly, these maintain a distinct condition of "active" chromatin that persists much later, between bouts of TCR-signaling, in the Th2 memory state. Thus, the active chromatin configuration around the locus is not simply an adjunct to active transcription nor dependent on ongoing NFAT recruitment to the promoter. The permissive state is associated not only with DNase hypersensitive sites but also with depletion of CpG methylation across the Th2 cytokine gene promoters, as well as the presence of "accessible" histone marks. This permissiveness presumably underlies the preferential recruitment of factors like NFAT to the *Il4* promoter in these cells, the next time TCR signaling is reactivated. The Th1 cytokine genes are not equivalently clustered, but the regulation of the "signature" Th1 cytokine gene *Ifng* has now been described in detail (Balasubramani et al. 2010; Schoenborn et al. 2007; Chang and Aune 2007). At the *Ifng* locus, an array of elements, extending to about 60 kb upstream and about 50 kb downstream of the transcription unit itself, turns out to be required for correct, fully efficient expression. Again, DNase hypersensitivity of these elements is greatly enhanced by Th1 differentiation and reduced by Th2 differentiation. Several of these key elements are directly engaged by the Th1 transcription factors T-bet (Tbx21) and Stat4 (Schoenborn et al. 2007; Wei et al. 2010; Balasubramani et al. 2010). These cases illustrate chromatin modifications that go along with lineage-specific memory and stable differentiation, and which can be separated from activation of pol II-dependent transcription per se.

5 Reversibility and Plasticity of Histone Marks

Nevertheless, histone marks are clearly subject to modification in response to changes in engagement of sequence-specific transcription factors [reviewed by (Natoli 2010)]. It is notable that across the genome the great majority of the "positive" marks, H3K4me2 and H3K4me3, are localized in discrete islands, with clear peaks implying a site-specific deposition mechanism (see examples in Fig. 1). Indeed, changes in histone marks can be a symptom of current

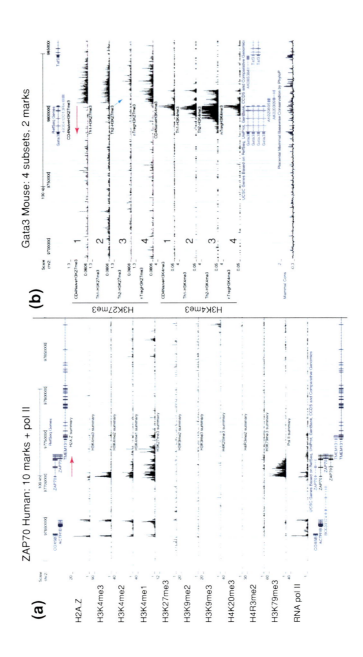

T-Cell Identity and Epigenetic Memory 125

◀ **Fig. 1** Examples of histone marking patterns in neighborhoods of strongly expressed T-cell genes. **a** *ZAP70*, in human CD4$^+$ T cells [data from (Barski et al. 2007)]. **b** *Gata3*, in four subsets of murine CD4$^+$ T cells: 1, naïve T cells; 2, Th1 T cells; 3, Th2 T cells, and 4, "natural" regulatory T cells (nTregs) [data from (Wei et al. 2009)]. Plots show enrichments of chromatin immune precipitation with antibodies against the indicated modified histones and RNA polymerase II, displayed as histograms against the human (hg18, panel a) and murine (mm9, panel b) genomes on the UCSC genome browser (http://genome.ucsc.edu). WIG files for **a** were downloaded from http://dir.nhlbi.nih.gov/papers/lmi/epigenomes/hgtcell.html. WIG files for **b** were constructed from raw data deposited in a public repository by Wei et al. and converted from mm8 to mm9 coordinates before plotting. H2A.Z, H3K4me1, H3K4me2, and H3K4me3 are all markers for accessibility and/or activation. H3K27me3, H3K9me2, H3K9me3, and H4K20me3 are all implicated in repression. H3K79me3 is associated with actively elongating polymerase. The repression mark H3K27me3 and the activation mark H3K4me3 are particularly useful, as shown in **b**. Magenta arrows indicate the genes of interest and their directions of transcription. In **b**, the blue arrow shows the site of upstream regulatory element for *Gata3* that is selectively demethylated in Th2 cells

transcription factor binding. For example, activation of the combination of factors that turns on *Il2* transcription can rapidly increase DNA accessibility to DNase and restriction endonuclease digestion (Ward et al. 1998; Rao et al. 2001) and demethylate CpG residues in the promoter-proximal region (Bruniquel and Schwartz 2003; Murayama et al. 2006). This coincides with de novo acetylation of histone H3 over several kilobases upstream of the promoter (Adachi and Rothenberg 2005; Chen et al. 2005).

Alteration of local histone modifications is a general correlate of the transcription factors at work. Binding of the crucial myeloid and B-cell factors PU.1 and EBF1 can rapidly induce mono- and/or di-methylation of H3K4, respectively, when they bind their target sites (Ghisletti et al. 2010; Heinz et al. 2010; Treiber et al. 2010). Nucleosome remodeling and local conversion of H3K4me1 into H3K4me2 or me3 marking, causing the H3K4me1 marked region to flatten and "spread", is also seen as a rapid, direct response to the local binding of E2A, EBF1, or PU.1 (Heinz et al. 2010; Lin et al. 2010).

Repressive marks as well as activation marks can be dynamic. The specific loss of "open" H3K4me3 marks can be triggered quickly by the withdrawal of Notch signaling, which causes the Notch-sensitive transcription factor RBP-Jκ (or CSL = CBF1, Suppressor of Hairless, Lag-1) to switch from binding a coactivator to binding a corepressor complex containing the H3K4me3-specific demethylase KDM5A (Jarid1a, RBP-2) (Liefke et al. 2010). Conversely, conversion of na T cells into Th2 cells involves elimination of long-standing H3K27me3 marks from an upstream regulatory region of *Gata3* (Fang et al. 2007; Amsen et al. 2007) (Fig. 1b). The transcription factor Tbx21 (T-bet) can directly strip H3K27me3 repressive marks from the promoters of some of its target genes, by recruitment of demethylases like Jmjd3 and Utx (Miller and Weinmann 2010). Finally, even the supposedly prohibitive modification of CpG DNA methylation has elegantly been shown to be removable by the combined action of transcription factors E2A, EBF1, and Runx1 in early B-cell development (Maier et al. 2004). Clearly, in order to promote these epigenetic modifications, the factors must have access to the DNA and the previous marks cannot be viewed as deterministic.

These considerations challenge the picture that histone modification patterns give past transcription factor action a lasting dominance over the scope of future transcription factor action. Such marks may not, for example, distinguish prohibited from temporarily quiet genes in a given cell type. Instead, they may offer access to a different, more focused, more developmentally interesting kind of insight. Histone marking patterns in a steady-state condition reflect the site-specific integrations of prior and current regulatory inputs. Thus, the location of *changes* in histone marking, across the time of a developmental transition, will be highly selective, and will reveal specific cis-elements where key transcription factors must be acting within that time frame to promote that change.

6 Combinatorial Transcription Factor Action and Epigenetic Modification

Genome-wide methods allow new light to be shed on the rules governing the reciprocal interaction of transcription factors with the epigenetic modification apparatus. One question is to evaluate globally how much transcription factor binding may really be limited by prior epigenetic modification, if that can be separated from the concurrent action of other transcription factors. Another is what determines the epigenetic consequences of a transcription factor's binding in a given cellular and DNA sequence context.

The foundational determinant of transcription factor binding occupancy is of course target site sequence recognition. Powerful high-throughput technologies for assessing binding by purified transcription factors in vitro have now provided a greatly improved measure of the quantitative preferences that particular transcription factors show for diverse site variants, when they are binding on their own (Berger and Bulyk 2009). However, the actual distribution of sites that transcription factors bind in vivo is not always congruent with the set of sites that they would be predicted to bind even based on these improved bioinformatic grounds. For most factors, only a subset of canonical sites are bound in vivo at any one time, and also there is significant recruitment to sites where no predicted target site exists (Badis et al. 2009; Gordân et al. 2009). These findings suggest that the sites where factors actually work in a given cell type can be affected positively by interaction with other factors, not just limited by site masking or through chromatin constraints.

Although transcription factors generally bind to many fewer sites than could be predicted on sequence grounds alone, the sites they bind are not simply predictable by degree of match to a measured position weight matrix. More importantly, they are cell type specific. A factor such as PU.1 is expressed at different levels in B lineage cells, myeloid lineage cells, and early T cells, but it binds to distinct patterns of sites in all three (Heinz et al. 2010) (Zhang et al. unpublished results). Although myeloid lineage cells express the highest amounts of PU.1, they fail to engage PU.1 at a number of genomic sites where it is bound in the much lower expressing B or early T-lineage cells (Heinz et al. 2010) (Zhang et al. unpublished results). This implies that classical mass-action is not the sole discriminant for site selection. In a similar way, E2A binding in pre-pro B cells occupies some sites that it will not occupy later in definitive pro-B cells (Lin et al. 2010), despite increasing overall expression of E2A at the later stage. Is the site discrimination in these cases caused by masking through repressive chromatin? To date, the only genome-wide assessment of chromatin "closing" can be provided by the mapping of repressive histone marks. In our own studies of early T-cell precursors, PU.1 site selectivity does not appear to result from masking by H3K27me3 marks in general (Zhang et al. unpublished results).

Instead, some contextual information that may be provided by other transcription factors determines where a factor will bind. One example is the developmental role of EBF1 in recruitment of E2A to a large number of new sites in early B cell precursors, sites where E2A can bind in close proximity to EBF1 and where new H3K4 methylation marks are formed (Lin et al. 2010). Another likely example is the selective binding and activity profile of EBF1 when ectopically expressed in a hematopoietic, as opposed to a nonhematopoietic, cellular context (Treiber et al. 2010). Finally, the B and myeloid lineage factor PU.1 is recruited to different spectra of binding sites in myeloid and B cell contexts, in large part because of its need to interact with C/EBP family factors in the former and with E2A and/or Oct family factors in the latter (Heinz et al. 2010). C/EBP factors and PU.1 can each recruit the other to bind to joint cis-elements. The implication is that a large fraction of binding is itself combinatorially defined.

Whether binding by interacting factors must be strictly coordinate or whether it can be mediated through a "pioneer" mechanism seems to vary according to the factors involved and the cis-regulatory sequences that are in play. When constitutively expressed transcription factors bind DNA, they may create a favored recruitment site for later-mobilized transcription factors, mediated through local histone modifications. For example, constitutively expressed PU.1 in myeloid cells can guide activation-dependent factors selectively to myeloid-specific response genes, very likely because it can induce permissive local H3K4 monomethylation and nucleosome remodeling at sites where it has bound (Natoli et al. 2011; Ghisletti et al. 2010; Heinz et al. 2010). Although the PU.1 binding is insufficient to turn on these activation-dependent target genes itself, it can focus the activity of the activation-dependent factors on a subset of their potential genome-wide targets for efficient and cell-type specific responses. In these cases, combinatorial transcription factor action is still important, but epigenetic changes allow different participants to be brought to the site asynchronously.

Although factors such as PU.1, E2A, and EBF1 often induce "accessibility" histone marks at their binding sites, for each of these factors there is also a minority of sites bound in vivo where these marks are induced barely if at all, even with strong transcription factor occupancy (Lin et al. 2010) (Zhang et al. unpublished results). Why can these factors induce H3K4 modification in some but not all cases? The combinatorial binding analyses of Lin et al. (2010) suggest that E2A binding alone has a different functional role from E2A binding together with cofactors such as FoxO1 and EBF1. Thus, factor-factor interactions beyond those needed to stabilize occupancy may explain how recruitment of Set7/9 and MLL-type H3K4 methyl transferases, which are thought to generate mono- and dimethyl H3K4 respectively (Allis et al. 2007), may fail to occur at some sites. A crucial open question is whether these H3K4-undermethylated sites are simply nonfunctional sites, or whether they reflect deployment of the transcription factor for a subset of their functional roles that may not involve "accessibility", such as chromosome looping or repression. There is a clear precedent from the activation and repression of different classes of target genes in erythroid cells by GATA-1, in collaboration with different partners and with the induction of different epigenetic marks (Yu et al. 2009; Fujiwara et al. 2009; Tripic et al. 2009; Cheng et al. 2009).

7 A Problem of Repression

The stability of repression is central to some of the most important questions about the inheritance of cellular identity. Cellular identity clearly involves prohibiting certain genes from being expressed, even while other genes can be on or off at different times, and it would be valuable to understand the biochemical distinction between permanent repression and conditional repression, if the mechanisms are actually different. For activating marks, the functional connection is easy to make. Acetylated histone H3 and H3K4 trimethylation not only are highly correlated with the sites of currently active promoters across the genome (Wang et al. 2008; Orford et al. 2008; Heintzman et al. 2007; Barski et al. 2007; Johnson et al. 2007; Roh et al. 2006); these are marks that are actively emplaced by the coactivator complexes recruited by positive regulatory factors, including p300/CBP, GCN5, and PCAF (KAT2A, B, and KAT3A,B), and MLL factors (KMT2 family), respectively, and they are then functionally implicated in the recruitment of RNA pol II cofactors [reviewed by (Allis et al. 2007; Kouzarides 2007)]. In contrast, H3K27me3 only indicates one type of Polycomb Repression Complex (PRC) 2-dependent repression. Other repressive marks have been described, but none are comprehensive (Barski et al. 2007). Those repression marks that are present may also be less sharply localized than activating marks. It is more frequent for an entire transcription unit or extended stretch of intergenic DNA to be associated with H3K27me3 marks than with H3 acetylation or H3K4 di- or trimethylation "activating/accessibility" marks.

When a transcription start site is modified by H3K27me3 and/or H3K9me3 in the absence of overt activating marks, then the gene is normally silent. However, this is not the only form of silencing that is found, in part because a gene can be silent due to a lack of local activators, interaction with a remote silencing element, or sequestration in a "silencing" compartment of the nucleus whether or not its promoter-proximal region has been modified with these marks. Filion et al. (2010) have shown that the genome of *Drosphila melanogaster* can be divided among five types of histone marked chromatin, where most of the genes that are transcriptionally silent are in domains that lack all known "repressive" as well as "activating" marks. This unmarked fraction is considerably larger in mammalian cells (Barski et al. 2007) (Zhang et al. unpublished results), where it includes a major fraction if not a majority of the silent loci. These "empty" genes, untranscribed but lacking both activating and known repressive marks, provide little insight into the mechanisms that have turned them off originally or keep them silent afterwards.

It is not clear whether all silent genes should require active repression or direct promoter blockade in order to stay silent. Many genes could remain silent passively, simply because the transcription factors that they require are themselves repressed. Interestingly, in our own studies, silent transcription factor genes are much more likely to bear H3K27 or H3 K9 trimethylation markings than the average for silent genes in the genome overall (Zhang, unpublished data). This may be a common feature of development from Drosophila to human, as an overwhelming majority of the PRC-repression target genes conserved between *D. melanogaster* and human appear to encode transcription factors (Schuettengruber et al. 2007). It is more likely for non-transcription factor target genes to remain silent without defined repressive marks.

It is likely that some transcriptional repressor proteins used in T-cell development work as a rule by specific mechanisms that bypass a need for PRC2-type complexes. In our own analyses of developing T-cell precursors (Zhang et al. unpublished results), repressive H3K27me3 marks are not present on the CD4 intronic silencer (Taniuchi et al. 2002) in DN cells, at stages when this silencer is acting, nor on the intergenic silencer between Rag1 and Rag2 (Yannoutsos et al. 2004) at stages when these genes are also repressed. The CD4 silencer seems not to be maintained by DNA methylation either (Zou et al. 2001). Since both CD4 and Rag1/Rag2 silencers are Runx dependent, it is possible that Runx factors as a rule use a repressive mechanism other than H3K27me3 or DNA methylation. Recent evidence suggests that Runx-mediated repression can involve interchromosomal interactions, which occur in particular nuclear membrane-associated compartments of the nucleus (Collins et al. 2011), possibly mediated by the known nuclear matrix binding activity of Runx factor C-terminal domains (Zeng et al. 1998). In any case, these exclusions mean that the absence of H3K27me3 (or H3K9me3) marks need not mean the absence of a repressive mechanism.

These findings make it difficult to use a single "snapshot" of a long-repressed gene to locate the regulatory element(s) through which the initial repression occurred. The power of epigenetic analysis will be greatly enhanced when diagnostic indicators for additional classes of repression can be discovered. However, valuable information can emerge nonetheless when time-resolved changes in repressively marked domains are detected.

8 Epigenetic Marking Events in T-Lineage Gene Activation from Stem-Cell Precursors

For T-cell development, the challenge is to explain the combination of persistence of central identity together with plasticity of functional characteristics. What kind of information about the programming of T-cell identity can we expect to obtain through mapping of epigenetic changes? If data are only available from mature T cells, the "activation" marks at annotated promoters are likely to agree with current transcriptional activity, and are not particularly helpful to explain the nature of the process that established this active gene expression. But we can ask whether and where distinctive epigenetic marks may have been deposited through the mechanisms that explain the long-term exclusion of non-T lineage properties. Also, looking beyond the transcription start sites themselves, we can use epigenetically marked regions to locate candidate sites for cis-regulatory elements of both T and non-T lineage genes. And most powerfully, if we can examine immature T cells, as well, to track the pattern of epigenetic marking across the genome through developmental time, we may be able to explain the separate mechanisms that operate during the distinctive staging of the T-cell specification and commitment process.

The origin of the "persistent" T-cell properties may itself have an interesting epigenetic component because of the way the T-cell program is activated in developing thymocytes. It requires many days of Notch pathway signaling, under the influence of Notch ligands in the thymus, in order to establish T-cell gene expression and lineage commitment [reviewed by (Rothenberg et al. 2008; Petrie and Zuniga-Pflucker 2007)]. However, afterwards the Notch signal becomes dispensable for $CD4^+$ and $CD8^+$ cell production (Wolfer et al. 2001; Tanigaki et al. 2004). Thus, even the onset of T cell identity raises the question of why such a sustained exposure to Notch pathway signals in the thymus is needed, persisting over days, in order to activate the T-cell program and eventually to block off other potential programs forever. The highly staged progression toward lineage commitment preceding TCR gene rearrangement would be consistent with a stepwise mechanism in which initial regulatory factor induction could lead to epigenetic changes that make cis-regulatory elements of a few target genes more accessible to additional regulatory factors, thus allowing the targets to be turned on or off once these additional factors are available [cf. references (Heinz et al. 2010; Hagman and Lukin 2005)]. If some target genes in these cases encode transcription factors themselves, then this process can be iterated several times, reaching more and more target genes as distinct sets of regulatory genes come into play. But is this true? At present, the published literature offers a few tantalizing clues.

Not all T-cell genes begin from an inaccessible state in stem and progenitor cells. In multipotent hematopoietic stem cells, low-level gene expression appears to proceed simultaneously at a range of loci that are ordinarily associated with mutually exclusive cell fates, a phenomenon known as multilineage priming (Ji et al. 2010; Ng et al. 2009; Weishaupt et al. 2010; Månsson et al. 2007; Pronk et al. 2007; Miyamoto et al. 2002; Hu et al. 1997). Some of the earliest hints for poised chromatin as a mechanism for multilineage priming in stem cells included DNase hypersensitivity of T-lineage specific elements in multipotent cells (Wotton et al. 1989). A recent study has now compared multiple histone marks and pol II occupancy patterns across the genome in hematopoietic stem cells, multipotent progenitors, and more restricted progenitors on the one hand and mature peripheral T cells on the other hand (Weishaupt et al. 2010). A number of genes with T-cell specific expression patterns and strong activation marks in mature T cells were found to be poised for activation with H3K4me3 marks in stem and progenitor cells long before transcription began, while others had their promoters bivalently marked (with H3K4me3 or H3Ac, and H3K9me3 or H3K27me3). Still other T-cell specific genes were apparently devoid of any marks in the stem and progenitor compartment (Weishaupt et al. 2010). The prevalence of poised genes and genes not obviously silenced raises the question of what trigger these genes are waiting for. However, this inference does not reveal the timing and ordering of epigenetic modification relative to gene activation.

T-cell gene expression is induced in several waves during the differentiation of pro-T cells both in vivo and in vitro (Yui et al. 2010; David-Fung et al. 2009; Tydell et al. 2007; Kawazu et al. 2007; Dik et al. 2005; Taghon et al. 2005; Tabrizifard et al. 2004). There is a general trend toward full mobilization of a "T-cell identity" gene expression program by the DN3 stage (Rothenberg et al. 2008), but no single "on switch" activates T-cell genes at a stroke. Thus there could be a role for epigenetic constraints that need to be removed stepwise at certain key loci. Our own studies are tracking the shifting patterns of epigenetic modification across the genome as they correlate with transcriptional changes at successive stages in this process (Zhang et al. unpublished results). To date, however, in the published literature, epigenetic changes during murine T-cell specification itself have been traced most closely at the level of changes in DNA CpG methylation (Ji et al. 2010). The major increase in T-cell gene expression in the DN2 and DN3 stages is marked by dramatic demethylation of CpG sites around genes involved in T-cell identity, including *Lck*, *Tcf7* (encoding TCF-1), and *Bcl11b* (Ji et al. 2010) (http://charm.jhmi.edu/hsc/). The demethylation of *Bcl11b* in particular appears tightly coupled with the activation of this highly T-lineage specific gene (Yui et al. 2010). Thus, the transcription factors activated during and immediately after the DN1 stage can specifically recruit demethylating enzymes to these DNA sequences, or else selectively block maintenance methylation of these sites during proliferation, opening up access for recognition by the next tier of DNA-binding proteins. This rich trove of evidence is likely to yield considerable insight into the gene regulation mechanisms in early T cells. However, there is still much to learn about the component processes that interlock to activate the T-cell program.

9 Epigenetic Repression of Non-T-Cell Genes During T-Lineage Commitment

T lineage commitment depends in principle on both positive and negative regulation. T-cell essential positive regulatory factors need to be activated and their expression stabilized to maintain T-cell identity gene regulation. In parallel, regulatory genes used in alternative pathways need to be repressed or kept permanently silent. Although effects on alternative-lineage cytokine receptor gene expression and homing receptor expression are also likely to be important for shaping the process of lineage restriction in vivo, as discussed elsewhere (Rothenberg 2011), a particularly central cell-intrinsic aspect is the silencing of transcription factors that could provide access to non-T cell developmental programs (Table 1). These alternative-fate regulatory genes in fact are likely to be silenced through several distinct waves of repression, because access to different non-T fates is lost at distinct stages in T-cell specification (Rothenberg 2011). One alternative, the NK cell program, remains closely linked to the T-cell program in regulatory terms and arguably continues to underlie mature cytolytic T cell function. Only one transcription factor yet described appears to distinguish NK cells from T cells, the zinc finger factor Zfp105 (Chambers et al. 2007; Li et al. 2010b) (http://www.immgen.org). But other fate alternatives have clearly distinct regulatory features that separate them from virtually all mature T cells. Thus, T-lineage commitment can be predicted to involve establishment of durable, robust repressive mechanisms at a specific set of important regulatory loci.

Loss of access to the B cell program should entail loss of inducibility of the crucial B-cell specific factors EBF1 and Pax5 (Mandel and Grosschedl 2010; Lukin et al. 2008; Cobaleda et al. 2007). Loss of access to the dendritic cell program is likely to involve loss of PU.1 (Carotta et al. 2010; Lefebvre et al. 2005), just as loss of access to other myeloid programs should entail loss of PU.1 and C/EBP family factors (Laiosa et al. 2006; Wölfler et al. 2010). Another program that must be repressed is the pluripotent stem/progenitor cell state itself, since pre-commitment T-cell precursors initially express a group of regulatory factors that are strongly implicated in stem and progenitor cell proliferative expansion or self-renewal. These factors include products of the known proto-oncogenes *Tal1* (SCL), *Lyl1*, *Hhex*, *Lmo2*, *Bcl11a*, *Erg*, and *HoxA* cluster genes as well as the Hox cofactor Meis1, PU.1 (*Sfpi1*) and Gfi1b (Yui et al. 2010; Rothenberg et al. 2010), all of which are implicated in stem and progenitor cell development and maintenance. Thus, silencing of these genes may play a role in the irreversibility of T-lineage choice. Expression of the stem/progenitor cell regulatory state is evidently fairly stable, as it persists into the DN2 stage when pro-T cells have visibly responded to Notch signaling and have already begun upregulating T cell identity genes. The stem/progenitor regulatory genes must then be actively repressed through a T-lineage specific mechanism, because in Bcl11b-deficient pro-T cells they continue to be expressed abnormally (Li et al. 2010a). Given this integral role in early T-cell precursors, it is particularly interesting to

T-Cell Identity and Epigenetic Memory

Table 1 Silencing of "Inappropriate" Genes in mature T-lineage cells

Alternative program	Stage fate excluded	Gene	Repression mark in mature T cells	Comments
Erythroid	Prethymic	*Gata1*	None	"Empty"
B cell	Thymic entry	*Ebf1*	H3K27me3, promoter > body	Silent in multipotent progenitors
B cell	Thymic entry	*Pax5*	H3K27me3, broad	Silent in multipotent progenitors
Myeloid	DN2	*Cebpa (C/EBPα)*	H3K27me3, bivalent	
Myeloid/progenitor	DN2-DN3	*Sfpi1 (SPI1, PU.1)*	None	"Empty"
Stem/progenitor	DN2-DN3	*Tal1 (SCL)*	H3K27me3, broad	
Stem/progenitor	DN2-DN3	*Lmo2*	H3K27me3, broad	No expression after DN1 stage
Stem/progenitor	DN2-DN3	*Gata2*	H3K27me3	No expression after DN1 stage
Stem/progenitor	DN2-DN3	*Hhex*	H3K27me3, broad	Bivalent in Th1, Th2
Stem/progenitor	DN2-DN3	*Gfi1b*	Light H3K27me3	
Stem/progenitor, B	DN2-DN3	*Bcl11a*	H3K27me3, mostly promoter-focused	Some expression until end of DN3 stage
NK	DN2-DN3	*Nfil3*	H3K27me3, promoter	Derepressed in Th1, Th2
NK	DN2-DN3	*Zfp105*	H3K27me3, broad	Gene silent in all known T cells
NK, CD8	DN2-DN3	*Eomes*	Bivalent	Silent in DN; not used in CD4 T
Immature T	β-selection (DN to CD4$^+$CD8$^+$)	*Hes1*	H3K27me3	Notch target, no expression after DN3
Immature T	Positive selection	*Rag1-Rag2*	Very light H3K27me3	
Immature T	Positive selection	*Dntt*	Very light H3K27me3; bivalent	No repression on promoter
CD8 T	Positive selection	*Cd8b1 (CD8B)*	No marks	Not used in CD4 T
CD8 T	Positive selection	*Cd8a*	H3K27me3	Not used in CD4 T

Timing of program exclusions taken from references reviewed in (Rothenberg 2011; Rothenberg et al. 2010), supplemented with gene expression data from J. Zhang et al. (unpublished data) and www.immgen.org. Data for modifications associated with genes in CD4$^+$ mature T cells were mined from (Weishaupt et al. 2010; Wei et al. 2009) for murine genomes and from (Barski et al. 2007) for human genomes. Raw reads for murine data were converted into WIG files, processed to convert mm8 coordinates into mm9 equivalents, and uploaded onto the UCSC Genome Browser (www.genome.ucsc.edu) for visualization relative to gene annotations. Some data on status of genes in multipotent prethymic progenitors were obtained from (Weishaupt et al. 2010). See Figs. 1 and 2 for examples consider how the stem/progenitor regulatory genes are eventually taken out of action.

Published data sets for mature T cells indicate that some of the alternative-lineage genes are indeed repressed through the imposition of epigenetic silencing marks (Fig. 2, Table 1). *Pax5* and *Ebf1* are heavily marked with H3K27me3 across their 5' ends and known regulatory elements in multiple effector subsets of mouse and human CD4$^+$ T cells (Weishaupt et al. 2010; Wei et al. 2009; Barski et al. 2007) (Table 1; e.g. Fig. 2c). The single NK-specific regulatory gene that is not expressed in any known subsets of T cells, namely Zfp105, is covered with repressive marks in mature T cells as well. Similarly, stem/progenitor-cell genes *Gata2*, *Tal1*, *Lmo2*, and *Bcl11a* have all acquired strong repression marks in mouse and human mature CD4$^+$ T cells (e.g. Fig. 2a, d). However, a role for alternative modes of repression is evident in the cases of PU.1 (*Sfpi1* gene) (Fig. 2b). PU.1 is silenced during T-lineage commitment in the DN2-DN3 transition, but is devoid of known repressive marks in mature CD4$^+$ T cells (Weishaupt et al. 2010; Wei et al. 2009; Barski et al. 2007). This is potentially associated with the ability of PU.1 to be "reawakened" during differentiation into the Th9 type of effector T cells, one response path for mature CD4$^+$ cells after antigen stimulation (Chang et al. 2010). However, Th9 cells constitute a rare subset, and for the overwhelming majority of T cells PU.1 remains permanently silent. Several other genes similarly appear to be silenced without strong repressive marks, and for these it is currently difficult to use epigenetic information to elucidate the timing and sites of action of T-lineage specific repressive mechanisms.

For the genes that are silenced with H3K27me3 deposition, two interesting questions arise. First, is the deposition of this mark an integral part of transcriptional silencing, or may it simply be allowed to occur once the actual transcriptional repression event has taken place? Can appearance of repressive marks follow afterwards, in the aftermath of developmental repression? For example, there are genes that have their promoters flanked by islands of strong H3K27me3 modification even when they are expressed, similar to the case of *Gata3* (Fig. 1b). In these cases, conceivably, the "open" state maintained at the promoter might depend on a continuous "antirepression" mechanism that limits the spread of this negative regulatory feature. If these genes stop being actively induced, it is possible that negative marks could be propagated across cis-elements of the gene simply as a result of prolonged absence of the activating complexes that previously kept them at bay.

Second, is the repression denoted by H3K27me3 marking truly permanent? In fact our own results (Zhang et al. unpublished) provide a number of cases where these marks can be removed rather quickly when developmentally important genes are activated. Furthermore, there are some surprising examples from the data of (Wei et al. 2009), where H3K27me3 marking appears to be partially reversible (Table 1). In these cases, genes that are fully repressed in thymocytes (Zhang et al. unpublished data) and naïve T cells alike may have their repression marks seemingly attenuated during antigen-dependent effector specialization in Th1 or Th2 cells, even if the genes are not re-expressed. It is interesting that T-cell

T-Cell Identity and Epigenetic Memory 135

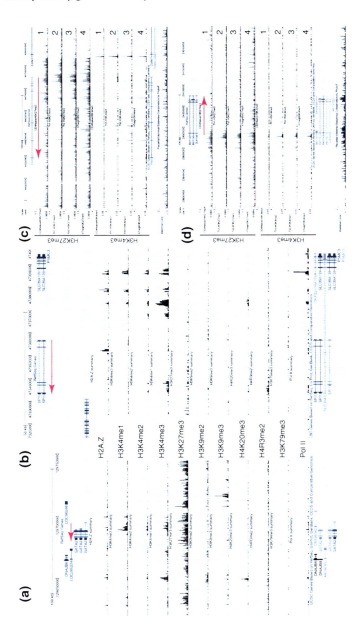

Fig. 2 Distinct modes of repression of lineage-inappropriate genes in T cells. **a** *Gata2* in human CD4⁺ T cells: an example of "broad" H3K27me3-mediated repression of a gene that was expressed in the earliest precursor stages. **b** *SPI1* in human CD4⁺ T cells: an example of repression without any repressive marks. Note the lack of any of the marks, H3K27me3, H3K9me2, H3K9me3, and H4K20me3, which are thought to mediate repression. **c** *Pax5* in four subsets of murine CD4⁺ T cells: an example of "broad" H3K27me3-mediated repression of a gene which may never have been expressed in T-cell precursors. **d** *Bcl11a* in four subsets of murine CD4⁺ T cells: an example of promoter-associated H3K27me3-mediated repression of a gene that is expressed in T cells until DN3 stage. In **c, d**: 1, naïve T cells; 2, Th1 T cells; 3, Th2 T cells, and 4, nTregs. Magenta arrows indicate the genes of interest and their directions of transcription. Data for **a, b** were from (Barski et al. 2007), as in Fig. 1a. Data for **c, d** were from (Wei et al. 2009), as in Fig. 1b

activation appears to result in the transfer of the histone H3K27 methyltransferase Ezh2 to the cytoplasm (Su et al. 2005; Hobert et al. 1996). If this indeed depletes the nuclear pool, then T cell activation could temporarily create a less restrictive chromatin environment for repressed genes with each round of DNA replication. Thus, something else besides the presence of the repressive marks defines their permanence in cases like Pax5, where no known reactivation mechanism appears to exist.

10 Concluding Remarks: Open Prospects

To explore how different regulatory mechanisms, durable and reversible ones, may be superimposed upon each other in T-cell development, epigenetic analysis offers a glorious harvest of detailed, gene-specific observations. However, it still presents some difficulties for explaining developmental programs. The information conveyed by specific histone and DNA methylation marks at a given moment in time is purely correlative, and so far, much of the use of this information has been to ratify what is already observed at the RNA expression level. For prediction or understanding of future gene activity and its constraints, there is still much to be learned. The H3K(9,14) Ac mark is tightly correlated with active promoters through its functional connection with RNA pol II recruitment, while the H3K4me3 modification marks both active and potentially active promoters (Heintzman et al. 2007). However, these give no clear way of distinguishing an inactive promoter that will be activated soon from one that was active but will never be active again. For repression, the H3K9me3 and H3K27me3 marks are highly correlated with silencing, but there are many silenced genes and possibly multiple silencing mechanisms that they miss. More poignantly, some genes that are covered with repressive marks and invested with CpG methylation at one stage may be strongly activated at a later stage, and we lack ways to distinguish prospectively the cases in which this is or is not possible.

Much interest has focused on the epigenetic status of known promoter regions that can be associated with known gene transcripts. Bioinformatically, due to rich

T-Cell Identity and Epigenetic Memory

genome annotation, this is the low-hanging fruit. However, the greatest payoff from epigenetic mapping is likely to come from the ability to visualize regulatory regions that are not part of known promoters or included in known transcription units. Some of these may actually be promoters of novel genes, including non-coding RNA genes, while others are distal cis-elements for known genes. In either case, the pattern of cis-elements that are in play is cell type specific and stage-specific, profoundly influenced by the way stage-specific transcription factors can affect histone modifications at their binding sites, as already described for PU.1, EBF1, and T-bet. Epigenetic marking patterns around developmentally regulated genes often suggest the presence of cis-elements that can be up to 100 kb away from the promoter or more, but visible because of histone modifications paralleling those at the promoter. Such elements may be crucial in order to eventually account for the regulation of the gene in various biological contexts.

With respect to T cells, a particularly interesting question is whether there is any link between different mechanisms of gene silencing and hierarchies of "fate exclusion" in development. The T-cell developmental pathway originates from hematopoietic committed cells in which lineage-irrelevant genes are permanently silent. Cells then appear to undergo an early exclusion of the erythroid and megakaryocytic developmental fates, followed by restrictions on myeloid fate, exclusion of the B cell fate, and then elimination of dendritic-cell and remaining myeloid potential together with loss of "stem/progenitor" properties (Rothenberg 2011). The NK-cell alternative may be eliminated last. One might imagine two extreme scenarios to account for these events. In the first, the depth of repression marking could be cumulative, so that non-hematopoietic and erythroid genes would be most heavily marked and NK cell genes the least heavily marked. Alternatively, at the opposite extreme, the heaviest repression marks might need to be applied in order to halt ongoing transcription, but eventually become dispensable as not only the gene itself but also the positive regulatory factors that once promoted its expression remain untranscribed. The incompleteness of our tools to detect different modes of repression limits the conclusions that can be drawn, but there are already data in hand to call into question both extreme models. The stem-cell genes *Lmo2* and *Gata2* carry repression marks in mature T cells, while the erythroid transcription factor gene *Gata1* remains in apparently open chromatin; yet H3K27me3 marks bury the B-cell specific *Pax5* and *Ebf1* genes, while the later-silenced PU.1 gene (*Sfpi1*) remains "open" (Table 1) (Barski et al. 2007; Wei et al. 2009; Weishaupt et al. 2010).

These results imply that distinct silencing marks may be less indicators of the "silent state" as a whole, or of genealogical hierarchy, than traces of the highly specific mechanism that acted to repress a given gene during a specific gene network transaction. In this way, they may be bringing us closer than we expected to the actual regulatory factors that are the causal forces in developmental lineage commitment.

The scope of evidence provided by genomic mapping of epigenetic marks in a given cell type is enormous, but the ability to use this evidence to answer specific developmental questions is enhanced and greatly focused when maps are compared from cells at successive stages of development. This follows from the goal of explaining changes in gene expression, which are necessarily derivatives of transcriptional activity with respect to time. The marked histone peaks that are relevant to the *change in expression* of a particular gene are most likely to be those that undergo a *change in modification* in parallel. Thus, rather than simply creating an impossibly large amount of data to analyze, sequential epigenetic analysis across a developmental process helps to narrow the field of focus to those particular transcription units and epigenetically marked regions that change status between stages. Cis-regulatory analysis of mammalian genes has not been as prominent in the past decade as it was earlier, partly because many strategies based on "Ockham's razor" logic have proven disappointing. However, by selecting the full set of candidate cis-elements identified as islands of dynamically specific epigenetic marking, and using modern gene transfer methods such as BAC transgenesis to assess their function, important regulatory questions should now come within range of solution.

Acknowledgments We gratefully acknowledge Georgi Marinov and Ali Mortazavi for help with software conversions between mm8 and mm9 mouse genomic coordinates, and members of the Rothenberg lab for stimulating discussions. The authors were supported by grants from the NIH, 5RC2 CA148278-02 and 3R01CA090233-08S1, and by the Albert Billings Ruddock Professorship (E.V.R.) at the California Institute of Technology.

References

Adachi S, Rothenberg EV (2005) Cell-type-specific epigenetic marking of the IL2 gene at a distal cis-regulatory region in competent, nontranscribing T-cells. Nucleic Acids Res 33:3200–3210

Allis CD, Berger SL, Cote J et al (2007) New nomenclature for chromatin-modifying enzymes. Cell 131:633–636

Amsen D, Antov A, Jankovic D et al (2007) Direct regulation of *Gata3* expression determines the T helper differentiation potential of Notch. Immunity 27:89–99

Amsen D, Spilianakis CG, Flavell RA (2009) How are T_H1 and T_H2 effector cells made? Curr Opin Immunol 21:153–160

Ansel KM, Djuretic I, Tanasa B et al (2006) Regulation of Th2 differentiation and *Il4* locus accessibility. Annu Rev Immunol 24:607–656

Araki Y, Fann M, Wersto R et al (2008) Histone acetylation facilitates rapid and robust memory CD8 T cell response through differential expression of effector molecules (eomesodermin and its targets: perforin and granzyme B). J Immunol 180:8102–8108

Araki Y, Wang Z, Zang C et al (2009) Genome-wide analysis of histone methylation reveals chromatin state-based regulation of gene transcription and function of memory $CD8^+$ T cells. Immunity 30:912–925

Badis G, Berger MF, Philippakis AA et al (2009) Diversity and complexity in DNA recognition by transcription factors. Science 324:1720–1723

Balasubramani A, Mukasa R, Hatton RD et al (2010) Regulation of the *Ifng* locus in the context of T-lineage specification and plasticity. Immunol Rev 238:216–232

Barski A, Cuddapah S, Cui K et al (2007) High-resolution profiling of histone methylations in the human genome. Cell 129:823–837

Berger MF, Bulyk ML (2009) Universal protein-binding microarrays for the comprehensive characterization of the DNA-binding specificities of transcription factors. Nat Protoc 4: 393–411

T-Cell Identity and Epigenetic Memory

Bernstein BE, Mikkelsen TS, Xie X et al (2006) A bivalent chromatin structure marks key developmental genes in embryonic stem cells. Cell 125:315–326

Bruniquel D, Schwartz RH (2003) Selective, stable demethylation of the interleukin-2 gene enhances transcription by an active process. Nat Immunol 4:235–240

Bunting K, Wang J, Shannon MF (2006) Control of interleukin-2 gene transcription: a paradigm for inducible, tissue-specific gene expression. Vitam Horm 74:105–145

Carotta S, Dakic A, D'Amico A et al (2010) The transcription factor PU.1 controls dendritic cell development and Flt3 cytokine receptor expression in a dose-dependent manner. Immunity 32:628–641

Chambers SM, Boles NC, Lin KY et al (2007) Hematopoietic fingerprints: an expression database of stem cells and their progeny. Cell Stem Cell 1:578–591

Chang S, Aune TM (2007) Dynamic changes in histone-methylation 'marks' across the locus encoding interferon-γ during the differentiation of T helper type 2 cells. Nat Immunol 8: 723–731

Chang HC, Sehra S, Goswami R et al (2010) The transcription factor PU.1 is required for the development of IL-9-producing T cells and allergic inflammation. Nat Immunol 11:527–534

Chen D, Rothenberg EV (1994) Interleukin-2 transcription factors as molecular targets of cAMP inhibition: delayed inhibition kinetics and combinatorial transcription roles. J Exp Med 179:931–942

Chen X, Wang J, Woltring D et al (2005) Histone dynamics on the interleukin-2 gene in response to T-cell activation. Mol Cell Biol 25:3209–3219

Cheng Y, Wu W, Kumar SA et al (2009) Erythroid GATA1 function revealed by genome-wide analysis of transcription factor occupancy, histone modifications, and mRNA expression. Genome Res 19:2172–2184

Chong MM, Simpson N, Ciofani M et al (2010) Epigenetic propagation of CD4 expression is established by the *Cd4* proximal enhancer in helper T cells. Genes Dev 24:659–669

Cobaleda C, Schebesta A, Delogu A et al (2007) Pax5: the guardian of B cell identity and function. Nat Immunol 8:463–470

Collins A, Hewitt SL, Chaumeil J et al (2011) RUNX transcription factor-mediated association of *Cd4* and *Cd8* enables coordinate gene regulation. Immunity 34:303–314

Cuddapah S, Barski A, Zhao K (2010) Epigenomics of T cell activation, differentiation, and memory. Curr Opin Immunol 22:341–347

Cui K, Zang C, Roh TY et al (2009) Chromatin signatures in multipotent human hematopoietic stem cells indicate the fate of bivalent genes during differentiation. Cell Stem Cell 4:80–93

Das R, Sant'Angelo DB, Nichols KE (2010) Transcriptional control of invariant NKT cell development. Immunol Rev 238:195–215

David-Fung E-S, Butler R, Buzi G et al (2009) Transcription factor expression dynamics of early T-lymphocyte specification and commitment. Dev Biol 325:444–467

Davidson EH (2006) The regulatory genome: gene regulatory networks in development and evolution. Academic Press, San Diego

Dik WA, Pike-Overzet K, Weerkamp F et al (2005) New insights on human T cell development by quantitative T cell receptor gene rearrangement studies and gene expression profiling. J Exp Med 201:1715–1723

Fang TC, Yashiro-Ohtani Y, Del Bianco C et al (2007) Notch directly regulates *Gata3* expression during T helper 2 cell differentiation. Immunity 27:100–110

Filion GJ, van Bemmel JG, Braunschweig U et al (2010) Systematic protein location mapping reveals five principal chromatin types in Drosophila cells. Cell 143:212–224

Fujiwara T, O'Geen H, Keles S et al (2009) Discovering hematopoietic mechanisms through genome-wide analysis of GATA factor chromatin occupancy. Mol Cell 36:667–681

Gangadharan D, Lambolez F, Attinger A et al (2006) Identification of pre- and postselection TCR$\alpha\beta^+$ intraepithelial lymphocyte precursors in the thymus. Immunity 25:631–641

Garrick D, Fiering S, Martin DIK et al (1998) Repeat-induced gene silencing in mammals. Nat Genet 18:56–59

Garrity PA, Chen D, Rothenberg EV et al (1994) IL-2 transcription is regulated in vivo at the level of coordinated binding of both constitutive and regulated factors. Mol Cell Biol 14:2159–2169

Ghisletti S, Barozzi I, Mietton F et al (2010) Identification and characterization of enhancers controlling the inflammatory gene expression program in macrophages. Immunity 32:317–328

Gordân R, Hartemink AJ, Bulyk ML (2009) Distinguishing direct versus indirect transcription factor-DNA interactions. Genome Res 19:2090–2100

Grimaud C, Negre N, Cavalli G (2006) From genetics to epigenetics: the tale of Polycomb group and Trithorax group genes. Chromosome Res 14:363–375

Hagman J, Lukin K (2005) Early B-cell factor 'pioneers' the way for B-cell development. Trends Immunol 26:455–461

Heintzman ND, Stuart RK, Hon G et al (2007) Distinct and predictive chromatin signatures of transcriptional promoters and enhancers in the human genome. Nat Genet 39:311–318

Heinz S, Benner C, Spann N et al (2010) Simple combinations of lineage-determining transcription factors prime cis-regulatory elements required for macrophage and B cell identities. Mol Cell 38:576–589

Heng TSP, Painter MW, Consortium TIGP (2008) The Immunological Genome Project: networks of gene expression in immune cells. Nat Immunol 9:1091–1094

Hobert O, Jallal B, Ullrich A (1996) Interaction of Vav with ENX-1, a putative transcriptional regulator of homeobox gene expression. Mol Cell Biol 16:3066–3073

Hu M, Krause D, Greaves M et al (1997) Multilineage gene expression precedes commitment in the hemopoietic system. Genes Dev 11:774–785

Jain J, Loh C, Rao A (1995) Transcriptional regulation of the IL2 gene. Curr Opin Immunol 7:333–342

Ji H, Ehrlich LI, Seita J et al (2010) Comprehensive methylome map of lineage commitment from haematopoietic progenitors. Nature 467:338–342

Johnson DS, Mortazavi A, Myers RM et al (2007) Genome-wide mapping of in vivo protein-DNA interactions. Science 316:1497–1502

Kawazu M, Yamamoto G, Yoshimi M et al (2007) Expression profiling of immature thymocytes revealed a novel homeobox gene that regulates double-negative thymocyte development. J Immunol 179:5335–5345

Kouzarides T (2007) Chromatin modifications and their function. Cell 128:693–705

Kreslavsky T, Savage AK, Hobbs R et al (2009) TCR-inducible PLZF transcription factor required for innate phenotype of a subset of $\gamma\delta$ T cells with restricted TCR diversity. Proc Natl Acad Sci USA 106:12453–12458

Laiosa CV, Stadtfeld M, Xie H et al (2006) Reprogramming of committed T cell progenitors to macrophages and dendritic cells by C/EBPα and PU.1 transcription factors. Immunity 25:731–744

Lee DU, Rao A (2004) Molecular analysis of a locus control region in the T helper 2 cytokine gene cluster: a target for STAT6 but not GATA3. Proc Natl Acad Sci USA 101:16010–16015

Lefebvre JM, Haks MC, Carleton MO et al (2005) Enforced expression of Spi-B reverses T lineage commitment and blocks β-selection. J Immunol 174:6184–6194

Li L, Leid M, Rothenberg EV (2010a) An early T cell lineage commitment checkpoint dependent on the transcription factor Bcl11b. Science 329:89–93

Li P, Burke S, Wang J et al (2010b) Reprogramming of T cells to natural killer-like cells upon Bcl11b deletion. Science 329:85–89

Liefke R, Oswald F, Alvarado C et al (2010) Histone demethylase KDM5A is an integral part of the core Notch-RBP-J repressor complex. Genes Dev 24:590–601

Lin YC, Jhunjhunwala S, Benner C et al (2010) A global network of transcription factors, involving E2A, EBF1 and Foxo1, that orchestrates B cell fate. Nat Immunol 11:635–643

Lukin K, Fields S, Hartley J et al (2008) Early B cell factor: Regulator of B lineage specification and commitment. Semin Immunol 20:221–227

Maier H, Colbert J, Fitzsimmons D et al (2003) Activation of the early B-cell-specific mb-1 ($Ig-\alpha$) gene by Pax-5 is dependent on an unmethylated Ets binding site. Mol Cell Biol 23:1946–1960

T-Cell Identity and Epigenetic Memory

Maier H, Ostraat R, Gao H et al (2004) Early B cell Factor cooperates with Runx1 and mediates epigenetic changes associated with mb-1 transcription. Nat Immunol 5:1069–1077

Mandel EM, Grosschedl R (2010) Transcription control of early B cell differentiation. Curr Opin Immunol 22:161–167

Månsson R, Hultquist A, Luc S et al (2007) Molecular evidence for hierarchical transcriptional lineage priming in fetal and adult stem cells and multipotent progenitors. Immunity 26:407–419

Meyer C, Zeng X, Chien YH (2010) Ligand recognition during thymic development and $\gamma\delta$ T cell function specification. Semin Immunol 22:207–213

Miller SA, Weinmann AS (2010) Molecular mechanisms by which T-bet regulates T-helper cell commitment. Immunol Rev 238:233–246

Miyamoto T, Iwasaki H, Reizis B et al (2002) Myeloid or lymphoid promiscuity as a critical step in hematopoietic lineage commitment. Dev Cell 3:137–147

Murayama A, Sakura K, Nakama M et al (2006) A specific CpG site demethylation in the human interleukin 2 gene promoter is an epigenetic memory. EMBO J 25:1081–1092

Murphy KM, Stockinger B (2010) Effector T cell plasticity: flexibility in the face of changing circumstances. Nat Immunol 11:674–680

Nakayama T, Yamashita M (2009) Critical role of the Polycomb and Trithorax complexes in the maintenance of CD4 T cell memory. Semin Immunol 21:78–83

Natoli G (2010) Maintaining cell identity through global control of genomic organization. Immunity 33:12–24

Natoli G, Ghisletti S, Barozzi I (2011) The genomic landscapes of inflammation. Genes Dev 25:101–106

Ng SY, Yoshida T, Zhang J et al (2009) Genome-wide lineage-specific transcriptional networks underscore Ikaros-dependent lymphoid priming in hematopoietic stem cells. Immunity 30:493–507

Onodera A, Yamashita M, Endo Y et al (2010) STAT6-mediated displacement of Polycomb by Trithorax complex establishes long-term maintenance of GATA3 expression in T helper type 2 cells. J Exp Med 207:2493–2506

Orford K, Kharchenko P, Lai W et al (2008) Differential H3K4 methylation identifies developmentally poised hematopoietic genes. Dev Cell 14:798–809

O'Shea JJ, Paul WE (2010) Mechanisms underlying lineage commitment and plasticity of helper CD4+ T cells. Science 327:1098–1102

Ouyang W, Löhning M, Gao Z et al (2000) Stat6-independent GATA-3 autoactivation directs IL-4-independent Th2 development and commitment. Immunity 12:27–37

Park K, He X, Lee HO et al (2010) TCR-mediated ThPOK induction promotes development of mature (CD24-) $\gamma\delta$ thymocytes. EMBO J 29:2329–2341

Petrie HT, Zuniga-Pflucker JC (2007) Zoned out: functional mapping of stromal signaling microenvironments in the thymus. Annu Rev Immunol 25:649–679

Polansky JK, Schreiber L, Thelemann C et al (2010) Methylation matters: binding of Ets-1 to the demethylated Foxp3 gene contributes to the stabilization of Foxp3 expression in regulatory T cells. J Mol Med 88:1029–1040

Pronk CJ, Rossi DJ, Mansson R et al (2007) Elucidation of the phenotypic, functional, and molecular topography of a myeloerythroid progenitor cell hierarchy. Cell Stem Cell 1:428–442

Rao S, Procko E, Shannon MF (2001) Chromatin remodeling, measured by a novel real-time polymerase chain reaction assay, across the proximal promoter region of the IL-2 gene. J Immunol 167:4494–4503

Reiner SL, Mullen AC, Hutchins AS et al (2003) Helper T cell differentiation and the problem of cellular inheritance. Immunol Res 27:463–468

Roh TY, Cuddapah S, Cui K et al (2006) The genomic landscape of histone modifications in human T cells. Proc Natl Acad Sci USA 103:15782–15787

Rothenberg EV (2011) T cell lineage commitment: identity and renunciation. J Immunol 186:6649–6655

Rothenberg EV, Ward SB (1996) A dynamic assembly of diverse transcription factors integrates activation and cell-type information for interleukin-2 gene regulation. Proc Natl Acad Sci USA 93:9358–9365

Rothenberg EV, Moore JE, Yui MA (2008) Launching the T-cell-lineage developmental programme. Nat Rev Immunol 8:9–21

Rothenberg EV, Zhang J, Li L (2010) Multilayered specification of the T-cell lineage fate. Immunol Rev 238:150–168

Schoenborn JR, Dorschner MO, Sekimata M et al (2007) Comprehensive epigenetic profiling identifies multiple distal regulatory elements directing transcription of the gene encoding interferon-γ. Nat Immunol 8:732–742

Schuettengruber B, Chourrout D, Vervoort M et al (2007) Genome regulation by Polycomb and Trithorax proteins. Cell 128:735–745

Schwartz YB, Pirrotta V (2007) Polycomb silencing mechanisms and the management of genomic programmes. Nat Rev Genet 8:9–22

Spits H, Di Santo JP (2011) The expanding family of innate lymphoid cells: regulators and effectors of immunity and tissue remodeling. Nat Immunol 12:21–27

Su IH, Dobenecker MW, Dickinson E et al (2005) Polycomb group protein Ezh2 controls actin polymerization and cell signaling. Cell 121:425–436

Tabrizifard S, Olaru A, Plotkin J et al (2004) Analysis of transcription factor expression during discrete stages of postnatal thymocyte differentiation. J Immunol 173:1094–1102

Taghon TN, David E-S, Zúñiga-Pflücker JC et al (2005) Delayed, asynchronous, and reversible T-lineage specification induced by Notch/Delta signaling. Genes Dev 19:965–978

Tanigaki K, Tsuji M, Yamamoto N et al (2004) Regulation of $\alpha\beta/\gamma\delta$ T cell lineage commitment and peripheral T cell responses by Notch/RBP-J signaling. Immunity 20:611–622

Taniuchi I, Sunshine MJ, Festenstein R et al (2002) Evidence for distinct *CD4* silencer functions at different stages of thymocyte differentiation. Mol Cell 10:1083–1096

Telfer JC, Hedblom EE, Anderson MK et al (2004) Localization of the domains in Runx transcription factors required for the repression of CD4 in thymocytes. J Immunol 172:4359–4370

Treiber T, Mandel EM, Pott S et al (2010) Early B cell Factor 1 regulates B cell gene networks by activation, repression, and transcription- independent poising of chromatin. Immunity 32:714–725

Tripic T, Deng W, Cheng Y et al (2009) SCL and associated proteins distinguish active from repressive GATA transcription factor complexes. Blood 113:2191–2201

Tydell CC, David-Fung E-S, Moore JE et al (2007) Molecular dissection of prethymic progenitor entry into the T lymphocyte developmental pathway. J Immunol 179:421–438

Wang Z, Zang C, Rosenfeld JA et al (2008) Combinatorial patterns of histone acetylations and methylations in the human genome. Nat Genet 40:897–903

Ward SB, Hernandez-Hoyos G, Chen F et al (1998) Chromatin remodeling of the interleukin-2 gene: distinct alterations in the proximal versus distal enhancer regions. Nucleic Acids Res 26:2923–2934

Wei G, Wei L, Zhu J et al (2009) Global mapping of H3K4me3 and H3K27me3 reveals specificity and plasticity in lineage fate determination of differentiating CD4+ T cells. Immunity 30:155–167

Wei L, Vahedi G, Sun HW et al (2010) Discrete roles of STAT4 and STAT6 transcription factors in tuning epigenetic modifications and transcription during T helper cell differentiation. Immunity 32:840–851

Weishaupt H, Sigvardsson M, Attema JL (2010) Epigenetic chromatin states uniquely define the developmental plasticity of murine hematopoietic stem cells. Blood 115:247–256

Williams A, Harker N, Ktistaki E et al (2008) Position effect variegation and imprinting of transgenes in lymphocytes. Nucleic Acids Res 36:2320–2329

Wilson CB, Rowell E, Sekimata M (2009) Epigenetic control of T-helper-cell differentiation. Nat Rev Immunol 9:91–105

Wolfer A, Bakker T, Wilson A et al (2001) Inactivation of Notch 1 in immature thymocytes does not perturb CD4 or CD8 T cell development. Nat Immunol 2:235–241

Wölfler A, Danen-van Oorschot AA, Haanstra JR et al (2010) Lineage-instructive function of C/EBPα in multipotent hematopoietic cells and early thymic progenitors. Blood 116:4116–4125

Wotton D, Flanagan BF, Owen MJ (1989) Chromatin configuration of the human CD2 gene locus during T-cell development. Proc Natl Acad Sci USA 86:4195–4199

Yamashita M, Shinnakasu R, Nigo Y et al (2004) IL-4-independent maintenance of histone modification of the IL-4 gene loci in memory Th2 cells. J Biol Chem 279:39454–39464

Yannoutsos N, Barreto V, Misulovin Z et al (2004) A cis element in the recombination activating gene locus regulates gene expression by counteracting a distant silencer. Nat Immunol 5:443–450

Yu M, Riva L, Xie H et al (2009) Insights into GATA-1-mediated gene activation versus repression via genome-wide chromatin occupancy analysis. Mol Cell 36:682–695

Yui MA, Feng N, Rothenberg EV (2010) Fine-scale staging of T cell lineage commitment in adult mouse thymus. J Immunol 185:284–293

Zeng C, McNeil S, Pockwinse S et al (1998) Intranuclear targeting of AML/CBFα regulatory factors to nuclear matrix-associated transcriptional domains. Proc Natl Acad Sci USA 95:1585–1589

Zhou L, Chong MM, Littman DR (2009) Plasticity of CD4+ T cell lineage differentiation. Immunity 30:646–655

Zhu J, Yamane H, Paul WE (2010) Differentiation of effector CD4 T cell populations. Annu Rev Immunol 28:445–489

Zou YR, Sunshine MJ, Taniuchi I et al (2001) Epigenetic silencing of CD4 in T cells committed to the cytotoxic lineage. Nat Genet 29:332–336

Encoding Stability Versus Flexibility: Lessons Learned From Examining Epigenetics in T Helper Cell Differentiation

Kenneth J. Oestreich and Amy S. Weinmann

Abstract It is currently unclear whether our classifications for T helper cell subtypes truly define stable lineages or rather they represent cells with a more flexible phenotype. This distinction is important for predicting the behavior of T helper cells during normal immune responses as well as in pathogenic conditions. Determining the mechanisms by which T helper cell lineage-defining transcription factors are expressed and subsequently regulate epigenetic and downstream gene regulatory events will provide insight into this complex question. Importantly, lineage-defining transcription factors that regulate epigenetic events have the potential to redefine the fate of the cell when they are expressed. In contrast, factors that regulate the events downstream of a permissive epigenetic environment will only have the capacity to modulate the underlying gene expression profile that is already established in that cell. Finally, mechanisms related to the antagonism versus cooperation between the lineage-defining factors for opposing T helper cell subsets will influence the characteristics of the cell. Here, we provide an overview of these topics by discussing epigenetic states in T helper cell subtypes as well as the mechanisms by which lineage-defining factors, such as T-bet, regulate gene expression profiles at both the epigenetic and general transcription level. We also examine some of what is known about the interplay between the T helper cell lineage-defining transcription factors T-bet, GATA3, Foxp3, Rorγt, and Bcl-6 and how this relates to the proper functioning of T helper cell subsets. Defining the mechanisms by which these factors regulate gene expression profiles will aid in our ability to predict the functional capabilities of T helper cell subsets.

K. J. Oestreich · A. S. Weinmann (✉)
Department of Immunology, University of Washington,
Box 357650,1959 NE Pacific Street, Seattle, WA 98195, USA
e-mail: weinmann@u.washington.edu

Current Topics in Microbiology and Immunology (2012) 356: 145–164
DOI: 10.1007/82_2011_141
© Springer-Verlag Berlin Heidelberg 2011
Published Online: 12 July 2011

Abbreviations

Bcl-6   B cell lymphoma-6
Foxp3   Forkhead box P3
H3K4   Histone 3 Lysine 4
H3K27   Histone 3 Lysine 27
HAT   Histone acetyltransferase
HDAC   Histone deacetylase
Jmjd3   Jumonji C domain-containing protein 3
RORγt   RAR-related orphan receptor gamma
T-bet   T-box expressed in T cells
Tfh   T follicular helper cell
Th   T helper
Treg   Regulatory T cell

Contents

1   T Helper Cell Lineages 146
2   Lineage-Defining Transcription Factors in Helper T Cell Development 147
3   Chromatin and Gene Expression 148
4   Chromatin Structure at Key Cytokine and Lineage-Defining Factor Loci 149
5   Simultaneous Expression of Lineage-Defining Factors: Friend or Foe? 152
6   The Role for T-bet in Th1 Cells 153
7   T-bet Regulates Prototypic Th1 Genes 153
8   T-bet Interacts with Chromatin-Modifying Complexes 154
9   Chromatin-Dependent Mechanisms for Other T Helper Cell
    Lineage-Defining Factors 156
10 Chromatin-Independent Activities for T-bet 157
11 Mechanisms for Antagonism Between Lineage-Defining Factors 157
12 Functional Cooperation Between Lineage-Defining Factors 159
13 Predicting the Functional Significance for the Co-Expression of Lineage-Defining
    Transcription Factors in T Helper Cells 160
References 161

1 T Helper Cell Lineages

It has long been recognized that naïve helper T cells can differentiate into specialized subsets that are designed to clear specific pathogens. T helper 1 (Th1) and Th2 cells were originally defined by their ability to stably produce IFNγ to coordinate the immune response to intracellular pathogens, or alternatively, IL-4 to coordinate the response to extracellular pathogens, respectively (Mosmann et al. 1986). More recently, however, it has become clear that T helper cells do not solely fit into the constraints of a model consisting of mutually exclusive Th1 versus Th2 lineages. We now know that several other T helper cell subtypes exist and contribute to the proper functioning of the immune response. Th17 cells are required for the clearance of extracellular bacteria and their dysregulation is now thought to play a pathogenic role in a number of autoimmune states (Langrish et al. 2005; Yen et al. 2006). Regulatory T cells (Treg) are required to keep immune responses in check, and the absence of these cells results in overwhelming systemic autoimmunity (Shevach 2009; Vignali et al. 2008). Another specialized subtype, T follicular helper cells (Tfh), is required for promoting B cell help and the loss of this subset severely impacts the generation of appropriate antibody responses (Crotty 2011; Yu and Vinuesa 2010). It also should be noted that several other T helper cell subtypes have recently been suggested based on cytokine expression profiles, but to date it is unclear if these are stable subtypes or rather represent a more transient state of a helper T cell (Murphy and Stockinger 2010; O'Shea and Paul 2010; Zhou et al. 2009). The phenotypes and importance of each of these helper T cell subsets has been expertly reviewed in depth elsewhere (Reinhardt et al. 2006; Zhou et al. 2009; Zhu and Paul 2008). Here, we will instead focus on the mechanisms by which lineage-defining transcription factors regulate the epigenetic states and transcriptional profiles of a helper T cell and how this relates to our concept of a stable lineage versus a more flexible subset.

2 Lineage-Defining Transcription Factors in Helper T Cell Development

The cytokines and chemokines that are secreted by innate immune cells in response to their initial encounter with a pathogen influence the ultimate identity of a helper T cell. This is because the environmental milieu that a naïve helper T cell encounters following activation regulates its differentiation by inducing specific cytokine-signaling pathways to promote the expression of unique lineage-defining transcription factors in that cell (O'Shea and Paul 2010; Zhu et al. 2010). Importantly, the combination of transcription factors present will then establish a specialized gene expression program in the T helper cell to create a functional subtype that is designed to coordinate the immune response to clear the encountered pathogen. For instance, innate immune cells will secrete cytokines such as IL-12 and IFNγ during the process of their initial encounter with an intracellular pathogen such as *Listeria monocytogenes*. Helper T cells that become activated in this environment will upregulate T-bet, destining them to express a Th1 gene expression profile. In contrast, a naïve helper T cell will upregulate GATA3 when they become activated in an environment where innate immune cells express IL-4. The expression of GATA3 will then contribute to the establishment of a Th2 gene expression profile. As illustrated by these two examples, the identity of the lineage-defining transcription factor(s) that are expressed will control the overall gene expression program and functional characteristics of the helper T cell.

Much research has been performed to identify and characterize the lineage-defining transcription factors that regulate T helper cell fate decisions. T-bet is required for Th1 cells, GATA3 for Th2 cells, RORγt for Th17 cells, Foxp3 for

Treg cells, and the transcriptional repressor Bcl-6 for Tfh cells (Hori et al. 2003; Ivanov et al. 2006; Johnston et al. 2009; Nurieva et al. 2009; O'Shea and Paul 2010; Szabo et al. 2000; Yu et al. 2009; Zheng and Flavell 1997). Importantly, if these factors are absent, the T helper cell subtype that they define will not develop. Therefore, the requirement for these factors is absolute. However, this is not to say that defining a helper T cell expression profile is as easy as determining whether one of these lineage-defining factors is present in the cell. In fact, it is now becoming clear that multiple T helper cell lineage-defining factors can be expressed in overlapping patterns, which may have profound consequences on the fate of the cell (Hegazy et al. 2010; Koch et al. 2009; Murphy and Stockinger 2010; Oestreich et al. 2011; Wei et al. 2009). Thus, one has to consider the totality of the transcription factors that are expressed in a given helper T cell as well as their functional capabilities in that context. This means that in order to predict the cellular profile, it is necessary to understand both the expression pattern and the mechanistic interplay between the lineage-defining transcription factors in a given cell.

3 Chromatin and Gene Expression

To understand the process by which T helper cells differentiate into functionally distinct subtypes, we will require detailed knowledge of the epigenetic differences between naïve helper T cells and each T helper cell functional subtype. Determining these epigenetic states, and the mechanisms by which they are established, will aid our predictions concerning the inherent stability or flexibility of T helper cell subsets. The chromatin or epigenetic state encompassing a gene and its regulatory regions provides an environmental and cell-type specific context to ensure that the expression of a particular gene is properly regulated in a given setting. This is because the DNA sequence for each gene is the same in every cell, but the "reading" of the DNA to allow for context-specific expression is different. Therefore, it is the chromatin or epigenetic environment for a gene that provides a means to package the DNA sequence so that it can be interpreted differently in unique cellular settings. Ultimately, the characteristics of the epigenetic environment directly influence the gene expression profile of an individual cell. There are a plethora of modifications that can occur on each histone, with the combination of these modifications representing one component of the epigenetic environment for a given gene.

Dynamic gene regulation, when a gene is rapidly induced from a resting state, is often associated with dramatic increases in histone acetylation (Jenuwein and Allis 2001). This is likely due to the recruitment of histone acetyltransfereases (HAT) in numerous co-activator and general transcription complexes. In contrast, developmental gene regulation is more often associated with changes in the composition of histone methylation states (Koyanagi et al. 2005; Mikkelsen et al. 2007; Ruthenburg et al. 2007; Shilatifard 2006). Historically, histone methylation has been thought to represent a completely stable epigenetic modification. However, with the identification of demethylase enzymes, it is becoming more appreciated that although there is stability to this modification, methylation can be developmentally regulated by specifically targeting the enzymatic machinery that creates and removes the methyl marks (Miller et al. 2008; Mosammaparast and Shi 2010; Shi 2007).

For the purposes of our discussion on helper T cell differentiation, we will place our focus on two well-studied histone methylation states, H3K4-methylation and H3K27-methylation, which are associated with permissive or repressive epigenetic states, respectively. It is also important to note that the co-localization of H3K4-methylation and H3K27-methylation, which has been termed bivalent chromatin, is suggested to represent a poised epigenetic state in progenitor cells for genes that are not yet fully expressed, but rather have the potential to be turned on or off in the subsequent developmental stage of the cell (Bernstein et al. 2006; Wei et al. 2009). To introduce some of our current understanding for lineage-specific epigenetic states in T helper cells, we will discuss research examining the chromatin structure at key cytokine loci as well as the loci of the lineage-defining transcription factors for T helper cells. In particular, extensive studies have analyzed the *Ifng* and Th2 cytokine loci due to their critical importance in modulating the immune response and the knowledge gained from these studies has also served as a general model for the role that epigenetic events play in developmental and inducible gene regulation. *Ifng* and Th2 cytokine gene regulation has been expertly reviewed elsewhere (Amsen et al. 2009; Schoenborn and Wilson 2007). Here, we will focus only on the epigenetic events that strictly correlate with their expression in specific T helper cell subsets as well as the role that the lineage-defining transcription factor T-bet plays in establishing the epigenetic environment at the *Ifng* locus.

4 Chromatin Structure at Key Cytokine and Lineage-Defining Factor Loci

One characteristic used to define T helper cell lineages is the expression of signature cytokines. Not surprisingly, each of the hallmark cytokine gene loci have a distinct epigenetic profile depending on whether they are expressed in a given T helper cell subtype (Ansel et al. 2003; Wilson et al. 2009). To illustrate this point, epigenetic profiling experiments have been performed in both a targeted fashion at cytokine loci as well as at the genome-wide level in multiple T helper cell lineages (Akimzhanov et al. 2007; Mukasa et al. 2010; Schoenborn et al. 2007; Wei et al. 2009; Zhou et al. 2004). These studies have in part contributed to the current paradigms correlating epigenetic states and the status of developmental gene expression. As an example, the *Ifng* locus is associated with permissive H3 and H4 lysine acetylation, as well as H3K4-methylation modifications in Th1 cells where it is expressed (Chang and Aune 2005; Hatton et al. 2006; Lewis et al. 2007; Schoenborn et al. 2007; Wei et al. 2009). In contrast, in naïve helper T cells or alternative T helper cell subtypes that do not express *Ifng*, the locus is associated with the repressive H3K27-methylation mark (Miller et al. 2008; Wei et al. 2009). Similar correlations between these epigenetic modifications and either an active or repressed gene expression state are also found at the Th2 cytokine loci and more recently at the *Il17a–Il17f* loci (Akimzhanov et al. 2007; Baguet and Bix 2004; Fields et al. 2004). Importantly, the epigenetic status of the gene in unique T helper cell subsets has strictly correlated with the T helper cell-restricted expression patterns for these cytokine loci.

Despite these paradigm "on-or-off" examples in which permissive and repressive epigenetic states are exclusively found in a canonical pattern correlating with T helper cell-specific cytokine expression profiles, recent data have also demonstrated that other categories of genes, in particular lineage-defining transcription factors, remain in a somewhat more plastic chromatin signature in multiple T helper cell subtypes (Wei et al. 2009). As mentioned above, a chromatin structure containing both permissive and repressive epigenetic marks is referred to as a bivalent state, which is thought to represent a poised, but not yet fully committed epigenetic status for that gene. For T helper cells, this may mean that these loci have the potential to be silenced or expressed. The decision to resolve the epigenetic state will be dependent on the signaling pathways that are upregulated in response to the cytokine milieu present. Importantly, both the *Tbx21* locus (which encodes T-bet) and the *Gata3* locus remain in a bivalent chromatin structure in multiple T helper cell subtypes suggesting that they retain the potential to be expressed longer than was previously appreciated (Wei et al. 2009) (Fig. 1). Thus, our prediction becomes that the stability of a T helper cell may be challenged if an environmental stimuli induces the expression of these factors in an opposing T helper cell lineage.

Determining the T helper cell-type specific bivalent status of the lineage-defining transcription factors helps us infer the potential for flexibility in T helper cell gene expression profiles. If the transcription factors for an opposing lineage retain the ability to be expressed, an inherent potential remains for that lineage to alter its underlying gene expression profile in response to environmental stimuli. Significantly, a bivalent chromatin structure can be found at the gene loci encoding the lineage-defining factors for Th1, Th2, and Th17 cells in the regulatory T cell population (Wei et al. 2009). This suggests that regulatory T cells have the potential to express a number of different lineage-defining factors depending upon the environmental conditions. Indeed, this epigenetic prediction has been observed with specialized regulatory T cell subsets that express T-bet or RORγt (Koch et al. 2009; Osorio et al. 2008). Importantly, the expression of these factors in regulatory T cells appears to create a specialized gene expression program that allows Treg cells to control pathogen-specific immune responses (Koch et al. 2009; Oldenhove et al. 2009).

The lack of a bivalent epigenetic state can also infer stable pathways that are less likely to be influenced in response to environmental cues after the T helper

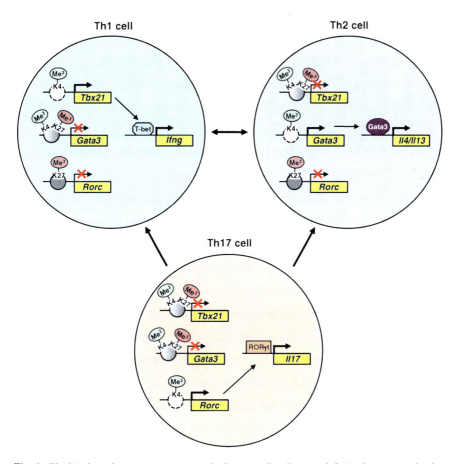

Fig. 1 *The bivalent chromatin structure at the loci encoding lineage-defining factors may lead to flexibility between T helper cell types.* This schematic displays three T helper cell types (Th1, Th2, and Th17). Inside each cell, the loci for the three lineage-defining factors are shown with an open chromatin structure (H3K4me3), a closed chromatin structure (H3K27me3), or a bivalent chromatin structure (both H3K4me3 and H3K27me3). In each cell type, high levels of H3K4me3 correspond to the expression of the lineage-defining factor for a given cell type (T-bet for Th1, GATA3 for Th2, and RORγt for Th17). In turn, the expression of this transcription factor then leads to the upregulation of the hallmark cytokine for each lineage. Interestingly, bivalent chromatin structure may allow for transitions between the T helper cell types indicated by the bi- or uni-directional arrows. Note that only a uni-directional arrow exists between Th17 and Th1 or Th2 cells because the *Rorc* locus is not associated with a bivalent chromatin structure in Th1 and Th2 cells, but is rather found in a strictly repressive chromatin state

cell subtype has developed (Murphy et al. 1996; Wei et al. 2009). For instance, the *Rorc* locus (which encodes RORγt) is found in a repressive epigenetic state in both Th1 and Th2 cells suggesting that it becomes permanently extinguished in these cell populations (Wei et al. 2009). Thus, a Th1 or Th2 cell will be less likely to express RORγt and have flexibility toward the Th17 phenotype (Fig. 1).

Collectively, the data generated examining these epigenetic patterns in T helper cells suggest that under specialized circumstances, certain lineage-defining factors may be expressed in a subset that was traditionally thought to be an opposing lineage, while others may be less dynamic and follow a more canonical view for their expression profile (Krawczyk et al. 2007; Wei et al. 2009). These epigenetic possibilities invoke an intriguing, and in some pathogenic cases, a potentially problematic scenario, in which specific combinations of lineage-defining transcription factors may be expressed in a single T helper cell type. Experiments addressing both the regulation and consequences for such scenarios are just starting to be performed, and as discussed below, it appears that the co-expression of lineage-defining transcription factors may be more common than once thought.

5 Simultaneous Expression of Lineage-Defining Factors: Friend or Foe?

To further support the epigenetic evidence that key transcription factors retain flexibility in their expression profile longer than previously appreciated, there are now a number of examples demonstrating the co-expression of two or more lineage-defining transcription factors in a single T helper cell subtype. For instance, studies have shown that the Th1 lineage-defining factor T-bet can be expressed in Tfh, Th17, and Treg subsets (Koch et al. 2009; Mukasa et al. 2010; Nurieva et al. 2009; Oldenhove et al. 2009; Wei et al. 2009). Furthermore, although Th2 cells were originally thought to be a fully differentiated stable cell lineage, a recent study identified a "Th2+1" cell, in which a Th2 cell upregulated T-bet expression in inflammatory conditions and gained many of the effector functions of a Th1 cell (Hegazy et al. 2010). Another example of this phenomenon is observed in the balance between the lineage-defining factors of the Th17 and Treg cell populations where the ability of RORγt to transactivate target genes is attenuated when it is co-expressed with Foxp3 (Zhou et al. 2008). The bivalent epigenetic state encompassing the lineage-defining transcription factors now provides a potential explanation for these observations because with appropriate environmental stimulation to alter the balance of the bivalent chromatin toward the more permissive state, the cell then will be able to express a second lineage-defining transcription factor because it was not permanently extinguished in repressive chromatin structure.

These observations now raise an important question. If multiple lineage-defining factors are expressed in a given T helper cell type, how will this impact the functional characteristics of the cell? To answer this question, we will need to determine how the expression (or co-expression) of lineage-defining transcription factors in a classically viewed opposing T helper cell subtype will alter the gene expression profile of the cell. This will require a detailed knowledge of the activities for each factor, both alone as well as in combination. Collectively, defining the mechanisms by which each transcription factor functionally regulates

Encoding Stability Versus Flexibility

target gene expression will allow us to predict the consequences for their over-lapping expression patterns in specialized scenarios during the immune response.

6 The Role for T-bet in Th1 Cells

The studies that have been performed determining the role for the lineage-defining transcription factor T-bet in Th1 cell differentiation illustrate the plethora of activities a single factor can perform and the value this knowledge can have for understanding the meaning behind its expression in a given cell population. T-bet is a T-box transcription factor that was originally identified due to its restricted expression in Th1 cells relative to Th2 cells (Szabo et al. 2000). T-box factors have long been known to play required roles in many developmental systems, and T-bet's role in T helper cell differentiation is no exception (Naiche et al. 2005). Indeed, experiments examining mice deficient in T-bet conclusively demonstrated its required role in Th1 cell differentiation (Finotto et al. 2002; Szabo et al. 2002). Interestingly, however, subsequent studies have shown that T-bet is not exclusively expressed in Th1 cells, but can also be expressed to varying degrees in several other T helper cell subtypes, as well as other immune cell types (Beima et al. 2006; Koch et al. 2009; Mukasa et al. 2010; Oldenhove et al. 2009). Significantly, its expression in these cell populations is functionally important, which raises the question of whether the T helper cell types that retain the ability to express T-bet after their initial differentiation will also have inherent flexibility in their phenotype. Before exploring the implications for these findings, we will first discuss what is currently known about the mechanisms by which T-bet is able to functionally establish gene expression profiles and how this knowledge impacts our view for the potential role T-bet is playing in these distinct cell types.

7 T-bet Regulates Prototypic Th1 Genes

Part of the role T-bet plays in establishing the Th1 cell phenotype is by activating the prototypic genes that classically define the characteristics of a Th1 cell. In particular, T-bet directly induces both *Ifng* and *Cxcr3* gene expression (Fig. 1). Several studies have shown that T-bet is able to associate with the *Ifng* and *Cxcr3* promoters in Th1 cells and that this interaction is required for their expression (Beima et al. 2006; Cho et al. 2003; Lewis et al. 2007; Lugo-Villarino et al. 2003; Szabo et al. 2002). Importantly, the follow-up studies examining the mechanisms by which T-bet regulates prototypic Th1 gene expression are shedding light on both the series of events that are required for the expression of these genes as well as the functional potential that T-bet possesses to more broadly regulate target genes in a given context (Miller and Weinmann 2010). This research, in combination with the epigenetic studies that have characterized the *Ifng* locus in helper

T cell development, are providing a strong basis to appreciate the events that are required for establishing key aspects of helper T cell lineage potential.

8 T-bet Interacts with Chromatin-Modifying Complexes

T-bet's ability to physically associate with and functionally recruit chromatin-modifying complexes to target genes has emerged as one important mechanism by which it regulates gene expression patterns (Lewis et al. 2007; Miller et al. 2008; Miller et al. 2010). This is particularly significant because it means that T-bet is not limited by the epigenetic environment present in the cell at the time it is expressed, but rather T-bet is involved in establishing new epigenetic states. Thus, T-bet, and other lineage-defining factors that are able to functionally modify the chromatin environment at their target genes have the potential to alter the inherent characteristics of a given cell upon their expression.

Significantly, T-bet is able to physically associate with several unique chromatin-modifying complexes, with each having distinct functional implications for regulating the expression of target genes (Fig. 2). T-bet has the potential to functionally reverse a repressive chromatin environment through its association with the H3K27-demethylase Jmjd3 (Miller et al. 2008; Miller et al. 2010). As mentioned previously, H3K27-methylation serves as a repressive epigenetic modification that silences genes during development and the cytokine genes that define unique T helper cell subsets are often found to have high levels of H3K27-methylation in both naïve helper T cells as well as the alternative T helper cell lineages that do not express the given cytokine (Wei et al. 2009). Thus, T-bet's ability to physically recruit an H3K27-demethylase containing complex to regulatory regions means that when T-bet is expressed, it has the fundamental ability to reverse the repressive chromatin landscape at its target genes in both naïve helper T cells as well as alternative T helper cell types. This means that instead of being limited by a repressive epigenetic state in a T helper cell lineage, T-bet has the capacity to change it.

T-bet is also able to physically interact with an H3K4-methyltransferase complex to functionally establish the permissive H3K4-methylation state at target genes (Lewis et al. 2007; Miller et al. 2008) (Fig. 2). Therefore, T-bet has the functional capacity to both reverse the repressive H3K27-methylation state of its target genes as well as establish a more permissive H3K4-methylation epigenetic state. Interestingly, T-bet's ability to associate with the H3K4-methyltransferase and H3K27-demethylase complexes requires physically separable residues in the T-box domain (Miller et al. 2008). This provides the potential for precise control of the epigenetic modifications present at a given target gene instead of a strictly all-or-nothing change from a repressive to a fully permissive epigenetic state. One can hypothesize that this may be important for creating gene-specific epigenetic patterns reflective of the need to finely tune the composition of the genes that are expressed in unique circumstances. Therefore, not all responses will necessarily

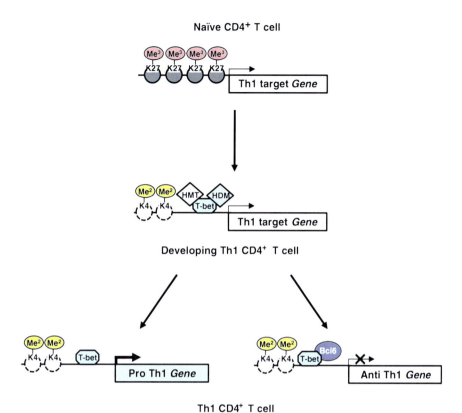

Fig. 2 *T-bet utilizes chromatin-dependent and chromatin-independent mechanisms to regulate gene expression and drive Th1 cell development.* In this schematic, a Th1 target gene is displayed initially with a high level of the repressive H3K27me3 modification. Following the initiation of Th1 development, T-bet associates with the locus, physically recruiting H3K27-demethylase (HDM) and H3K4-methyltransferase (HMT) complexes to functionally remove the H3K27me3 mark, while adding the H3K4me2 mark, respectively. The resulting accessible chromatin structure allows for either gene activation via T-bet-dependent transactivation in the case of a gene expressed in Th1 cells (pro-Th1 gene), or repression via the T-bet-dependent recruitment of the transcriptional repressor Bcl-6 (anti-Th1 gene). In this manner, T-bet can direct Th1 lineage commitment by both upregulating prototypic Th1 gene targets and repressing genes associated with the formation of alternative T helper cell types

result in identical changes to the gene expression profile, and ultimately functional characteristics, of the cell.

In addition to creating a more permissive epigenetic state through modifying the status of histone methylation, T-bet also functionally induces a broader general remodeling of the chromatin at target genes. This is because Jmjd3, the H3K27-demethylase that complexes with T-bet, also physically associates with a Brg1-containing SWI/SNF-remodeling complex in a demethylase-independent manner (Miller et al. 2010). Therefore, the interaction between T-bet and Jmjd3 has the capacity to mediate both the transition from the repressive H3K27-methylation state as well as a general opening of the chromatin that can effectively uncover binding sites for other regulatory factors. Once again, the separable nature of Jmjd3's H3K27-demethylase activity versus its ability to physically associate with the SWI/SNF chromatin-remodeling complex provides flexibility to perform these functions in a context-specific fashion. One hypothesis is that the H3K27-demethylase activity will be more prominent during developmental transitions, while the general chromatin-remodeling activity will play a greater role in terminally differentiated cell types. Collectively, in building a mechanistic hierarchy to logically predict the effects for the expression of the lineage-defining transcription factors, T-bet expression has the potential to induce widespread, yet highly specific, alterations in the epigenetic and general chromatin landscape of the cell.

9 Chromatin-Dependent Mechanisms for Other T Helper Cell Lineage-Defining Factors

In addition to the in-depth studies determining the mechanisms by which T-bet modulates the epigenetic environment, research examining other T helper cell lineage-defining factors has also been performed. A number of elegant experiments have correlated the expression of various T helper cell lineage-defining factors with changes in chromatin structure at key cytokine loci and have identified some of their interacting partner proteins. For example, during Th2 differentiation, the GATA3-dependent activation of the Th2 cytokine loci is accompanied by the induction of permissive histone modifications (Baguet and Bix 2004; Fields et al. 2002; Lee et al. 2006). Significantly, these epigenetic changes are dependent upon the presence of GATA3 (Ansel et al. 2006; Lee et al. 2001; Lee et al. 2000). The exact mechanism by which GATA3 regulates permissive histone modifications is still being explored. Additionally, other studies have shown that GATA3 may also play a role in regulating DNA methylation (Hutchins et al. 2002; Makar et al. 2003). Like histone modifications, DNA methylation represents another important epigenetic event that influences the ability of transcription factors to access their DNA binding elements in a given cell (Wilson et al. 2009). Taken together, it is clear that GATA3 also plays a role in functionally regulating epigenetic states.

Genome-wide studies examining Foxp3 binding patterns have provided strong evidence that Foxp3 inherently functions as a T helper cell lineage-defining transcription factor by both activating and repressing gene expression in regulatory T cells (Zheng et al. 2007). In part, Foxp3 appears to utilize epigenetic mechanisms to regulate its target genes. Importantly, Foxp3 has been found in multiprotein complexes that include chromatin-remodeling proteins such as Brg1 and Mbd3 (Li and Greene 2007). In addition, recent data suggest that Foxp3 may also functionally regulate the chromatin structure and expression of a subset of target genes through its ability to interact with other transcription factors such as STAT3 and Eos (Chaudhry et al. 2009; Pan et al. 2009). Collectively, the research performed to date suggests that the T helper cell lineage-defining transcription factors T-bet, GATA3, and Foxp3 function at least in part by regulating the epigenetic environment of the cells in which they are expressed. Experiments to identify the molecular mechanisms utilized by the other lineage-defining transcription factors in T helper cells are just beginning and should yield important insights into their functional capabilities as well.

10 Chromatin-Independent Activities for T-bet

In addition to regulating the epigenetic environment, lineage-defining factors also influence gene expression events that occur subsequent to chromatin remodeling. Here again, studies examining T-bet illustrate this point. Significantly, T-bet's ability to create a permissive chromatin environment is required but not sufficient to activate endogenous *Ifng* and *Cxcr3* gene expression (Lewis et al. 2007; Miller et al. 2008). These data have indicated that T-bet possess a general transactivation potential that is independent from its chromatin-remodeling functions and this activity is required for the upregulation of at least a subset of target genes. The exact mechanism(s) for this activity are currently unknown, but it demonstrates the importance for the series of events that occur after a permissive epigenetic state has been established to determine the final gene expression outcome. This point is further illustrated by recent findings elucidating a mechanism by which T-bet can functionally repress a subset of target genes that are important for regulating alternative T helper cell lineages (Oestreich et al. 2011). In this case, T-bet physically associates with the transcriptional repressor Bcl-6 and recruits it to a subset of target promoters to downregulate gene expression in Th1 cells (Oestreich et al. 2011) (Fig. 2). This interaction effectively converts T-bet into a site-specific transcriptional repressor in some circumstances. The association between T-bet and Bcl-6 also highlights the potential for creating unique gene expression profiles when two lineage-defining transcription factors are expressed in the same cell (Fig. 3). Next, we will discuss mechanisms by which the co-expression of classically viewed opposing lineage-defining transcription factors can impact gene expression in T helper cell subsets.

11 Mechanisms for Antagonism Between Lineage-Defining Factors

The classical view for the mechanisms by which opposing T helper cell lineage-defining factors regulate each other's functional activities is through antagonism. Numerous studies have demonstrated that the T helper cell lineage-defining factors can interfere with both the expression level and functional capability of the

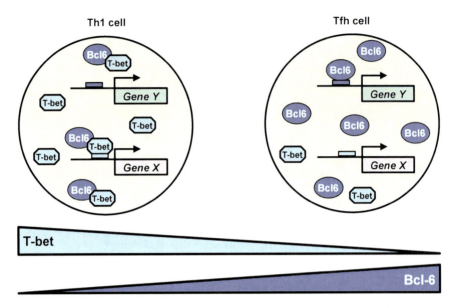

Fig. 3 *The balanced expression of lineage-defining transcription factors may determine the identity of a T helper cell.* In this schematic, a Th1 versus Tfh cell are shown. T-bet is in excess relative to Bcl-6 in a Th1 cell, while the opposite expression pattern is found in the Tfh cell. The high expression levels of T-bet allow it to physically associate with and recruit Bcl-6 to a subset of T-bet target genes in Th1 cells. Importantly, because Bcl-6 is limiting, its association with T-bet prevents it from binding to its own DNA-binding elements. In contrast, in the Tfh cell, the overall abundance of Bcl-6 ensures that it can associate with its own target genes, creating a Tfh cell-specific gene expression profile. One can imagine similar scenarios for the co-expression of other T helper cell lineage-defining factors

opposing factor in order to promote the differentiation of a single T helper cell type (Hwang et al. 2005; Lazarevic et al. 2011; Zhou et al. 2008). At the level of expression, often times the factors contribute, either directly or indirectly, to the repression of the opposing lineage-defining factors' gene transcription. For instance, the Tfh factor Bcl-6 has been suggested to directly repress the expression of the alternative T helper cell lineage-defining factors (Nurieva et al. 2009; Yu et al. 2009). Recently, an indirect mechanism by which T-bet antagonizes the expression of *Rorc*, the lineage-defining factor for Th17 differentiation, was uncovered (Lazarevic et al. 2011). In this case, T-bet interacts with Runx1 and prevents it from positively regulating the *Rorc* locus. Thus, both direct and indirect repression mechanisms contribute to the downregulation of the gene expression for the alternative transcription factor.

At the level of protein complex formation, inhibition mechanisms that are both physical and functional in nature have been identified. The lineage-defining factors T-bet and GATA3 have been shown to form a physical complex that impairs Th2 cell formation (Hwang et al. 2005). This interaction leads to the inactivation of the Th2 gene expression program by blocking the association of GATA3 with its target genes. Similarly, an interaction between Foxp3 and RORγt, the Treg and Th17 cell lineage-defining factors, respectively, has been shown to preferentially favor Treg formation by inhibiting the RORγt-dependent activation of IL-17 (Zhou et al. 2008). In both of these scenarios, the lineage-defining factor that is in excess may effectively sequester the opposing factor from its target genes. Therefore, the inherent functional activities of the transcription factors are not altered, but rather the limiting factor is not appropriately targeted to their classic binding sites.

12 Functional Cooperation Between Lineage-Defining Factors

Despite these well-known cases in which T helper cell lineage-defining transcription factors oppose one another, emerging research now suggests that perhaps in some circumstances they may also work together to promote the proper functioning of a given T helper cell type. Conceptually, this changes the way that we have sometimes simplistically viewed the expression of a lineage-defining transcription factor in a T helper cell as evidence that the cell must then have certain characteristics. Instead, it is now becoming clear that the overall composition of the factors present in a given epigenetic background will ultimately be responsible for determining the functional characteristics of the cell. This is why elucidating the mechanisms by which each factor regulates gene expression, both alone and in combination, is so important because it will give us the ability to predict the functional consequence for the co-expression of T helper cell lineage-defining factors in normal as well as pathogenic states.

The demonstration that the Th1 lineage-defining factor T-bet works together with the Tfh lineage-defining factor Bcl-6 to establish a Th1 signature gene expression profile is a good example to illustrate this point (Oestreich et al. 2011). Importantly, the interaction between T-bet and Bcl-6 inherently alters the functional capacity of T-bet and allows it to directly repress a subset of target genes in Th1 cells (Oestreich et al. 2011) (Fig. 2). Part of the mechanism that allows for an effective collaboration between two opposing lineage-defining transcription factors must be in the balance between their expression levels in different cellular settings. For instance, in Th1 cells, T-bet is expressed at high levels while Bcl-6 is expressed at comparatively low levels (Crotty et al. 2010; Oestreich et al. 2011). In Tfh cells, the opposite expression profile is found with high levels of Bcl-6 and relatively low levels of T-bet (Johnston et al. 2009; Nurieva et al. 2009; Yu et al. 2009). Therefore, one model for the mechanism by which the presence of Bcl-6 is needed to promote a Th1 gene expression profile without tipping the balance of the cell toward a Tfh phenotype is that the high T-bet to Bcl-6 ratio causes the majority of Bcl-6 to form a complex with T-bet, leaving little free Bcl-6 available (Fig. 3). This means that the majority of the Bcl-6 found in a Th1 cell is recruited by T-bet to its target genes, and because T-bet is in excess, there still is a significant amount of T-bet that is targeted to gene loci without Bcl-6 in complex. This allows T-bet to simultaneously direct both gene activation and repression programs in a Th1 cell. Although this is a simplified scenario (because both T-bet and Bcl-6 have numerous protein interaction partners), it serves to illustrate how the balance between the expression levels for normally opposing lineage-defining transcription factors may contribute to the regulation of complex, yet highly specific, gene expression profiles. Taken together, the balance of the factors present could promote the T-bet-dependent recruitment of chromatin-remodeling complexes, activating complexes, and Bcl-6 repressive complexes to subsets of target promoters in Th1 cells, while inducing a different series of events in Tfh cells or alternative T helper cell subsets.

13 Predicting the Functional Significance for the Co-Expression of Lineage-Defining Transcription Factors in T Helper Cells

It has now become clear that T helper cell lineage-defining transcription factors can be found in overlapping patterns and these expression patterns can have variable consequences dependent upon the nuclear and epigenetic environment of the cell. This topic has significant implications for understanding the inherent stability or flexibility for T helper cell subsets. Mechanistically, we will need to understand the functional capabilities of each factor to predict the possible consequences for their expression in specific cellular settings. We will also need to know the global epigenetic background of the cell. The transcription factors that are able to influence the chromatin environment will be at the top of a mechanistic hierarchy because they are not limited by the current epigenetic state of the cell. Thus, if a T helper cell lineage-defining factor remains in a bivalent epigenetic state and an environmental stimulus induces its expression in an alternative lineage, it will have the ability to dramatically alter the characteristics of the cell because it is not limited to modulating only the genes that are currently in a permissive epigenetic state in that T helper cell type. In contrast, the transcription factors that are only able to associate with regulatory regions that are already found within an accessible chromatin structure are inherently more limited in their potential. Specifically, they will only have the ability to modulate the current gene expression program, but not intrinsically change the underlying profile. Therefore, these transcription factors will be further down the hierarchy because they do not control their own "fate" in regard to which target genes they can regulate in a given cell, but rather the epigenetic environment of the cell directs their potential.

In addition to assessing the role for the lineage-defining factors in regard to the epigenetic environment of the cell, it will also be important to determine the physical interactions between the factors. Significantly, physical interactions will have the potential to alter the functional capabilities of one or both of the factors in a given setting. In some cases, this may mean that the simultaneous expression of two factors will impact the functional capacity of each transcription factor.

We observe this with the association between T-bet and Bcl-6, where this interaction allows T-bet to effectively target a specific gene repression program in addition to its more classical role as a transcriptional activator. Therefore, the results for the combinatorial expression of two or more lineage-defining factors in T helper cells will not be as simple as a sum of the individual activities.

The complex nature by which T helper cell-specific gene expression programs are established prevents us from simply pooling all of our mechanistic data for the lineage-defining transcription factors and predicting a final outcome on the fate of the cell. However, by defining the epigenetic states in T helper cell subsets and the mechanisms by which they are established and maintained, we come a step closer to understanding the inherent stability or flexibility of T helper cells. The regulation of the events that occur downstream of the epigenetic environment also represent an important area of control because the interplay between transcription factors will modulate the underlying gene expression programs established in a given cellular environment. As we uncover the molecular events that are regulated by each T helper cell lineage-defining transcription factor, it becomes more realistic to achieve a goal of logically predicting the potential that the expression of a specific factor will have in a given context. This will aid in our ability to assess possible functional consequences on T helper cell activity in normal and pathogenic settings.

Acknowledgments We thank members of the Weinmann lab for helpful discussions. The research performed in the authors' lab is supported by grants from the NIAID (AI061061) and (AI07272-061A) and the American Cancer Society (RSG-09-045-01-DDC).

References

Akimzhanov AM, Yang XO, Dong C (2007) Chromatin remodeling of interleukin-17 (IL-17)-IL-17F cytokine gene locus during inflammatory helper T cell differentiation. J Biol Chem 282:5969–5972

Amsen D, Spilianakis CG, Flavell RA (2009) How are T(H)1 and T(H)2 effector cells made? Curr Opin Immunol 21:153–160

Ansel KM, Lee DU, Rao A (2003) An epigenetic view of helper T cell differentiation. Nat Immunol 4:616–623

Ansel KM, Djuretic I, Tanasa B, Rao A (2006) Regulation of Th2 differentiation and Il4 locus accessibility. Annu Rev Immunol 24:607–656

Baguet A, Bix M (2004) Chromatin landscape dynamics of the Il4-Il13 locus during T helper 1 and 2 development. Proc Natl Acad Sci USA 101:11410–11415

Beima KM, Miazgowicz MM, Lewis MD, Yan PS, Huang TH, Weinmann AS (2006) T-bet binding to newly identified target gene promoters is cell type-independent but results in variable context-dependent functional effects. J Biol Chem 281:11992–12000

Bernstein BE, Mikkelsen TS, Xie X, Kamal M, Huebert DJ, Cuff J, Fry B, Meissner A, Wernig M, Plath K et al (2006) A bivalent chromatin structure marks key developmental genes in embryonic stem cells. Cell 125:315–326

Chang S, Aune TM (2005) Histone hyperacetylated domains across the Ifng gene region in natural killer cells and T cells. Proc Natl Acad Sci USA 102:17095–17100

Chaudhry A, Rudra D, Treuting P, Samstein RM, Liang Y, Kas A, Rudensky AY (2009) CD4+ regulatory T cells control TH17 responses in a Stat3-dependent manner. Science 326:986–991

Cho JY, Grigura V, Murphy TL, Murphy K (2003) Identification of cooperative monomeric Brachyury sites conferring T-bet responsiveness to the proximal IFN-gamma promoter. Int Immunol 15:1149–1160

Crotty S (2011) Follicular helper CD4 T Cells (T(FH)). Annu Rev Immunol 29:621–663

Crotty S, Johnston RJ, Schoenberger SP (2010) Effectors and memories: Bcl-6 and Blimp-1 in T and B lymphocyte differentiation. Nat Immunol 11:114–120

Fields PE, Kim ST, Flavell RA (2002) Cutting edge: changes in histone acetylation at the IL-4 and IFN-gamma loci accompany Th1/Th2 differentiation. J Immunol 169:647–650

Fields PE, Lee GR, Kim ST, Bartsevich VV, Flavell RA (2004) Th2-specific chromatin remodeling and enhancer activity in the Th2 cytokine locus control region. Immunity 21:865–876

Finotto S, Neurath MF, Glickman JN, Qin S, Lehr HA, Green FH, Ackerman K, Haley K, Galle PR, Szabo SJ et al (2002) Development of spontaneous airway changes consistent with human asthma in mice lacking T-bet. Science 295:336–338

Hatton RD, Harrington LE, Luther RJ, Wakefield T, Janowski KM, Oliver JR, Lallone RL, Murphy KM, Weaver CT (2006) A distal conserved sequence element controls IFNg gene expression by T cells and NK cells. Immunity 25:717–729

Hegazy AN, Peine M, Helmstetter C, Panse I, Frohlich A, Bergthaler A, Flatz L, Pinschewer DD, Radbruch A, Lohning M (2010) Interferons direct Th2 cell reprogramming to generate a stable GATA-3(+)T-bet(+) cell subset with combined Th2 and Th1 cell functions. Immunity 32:116–128

Hori S, Nomura T, Sakaguchi S (2003) Control of regulatory T cell development by the transcription factor Foxp3. Science 299:1057–1061

Hutchins AS, Mullen AC, Lee HW, Sykes KJ, High FA, Hendrich BD, Bird AP, Reiner SL (2002) Gene silencing quantitatively controls the function of a developmental trans-activator. Mol Cell 10:81–91

Hwang ES, Szabo SJ, Schwartzberg PL, Glimcher LH (2005) T helper cell fate specified by kinase-mediated interaction of T-bet with GATA-3. Science 307:430–433

Ivanov II, McKenzie BS, Zhou L, Tadokoro CE, Lepelley A, Lafaille JJ, Cua DJ, Littman DR (2006) The orphan nuclear receptor RORgammat directs the differentiation program of proinflammatory IL-17+ T helper cells. Cell 126:1121–1133

Jenuwein T, Allis CD (2001) Translating the histone code. Science 293:1074–1080

Johnston RJ, Poholek AC, DiToro D, Yusuf I, Eto D, Barnett B, Dent AL, Craft J, Crotty S (2009) Bcl6 and Blimp-1 are reciprocal and antagonistic regulators of T follicular helper cell differentiation. Science 325:1006–1010

Koch MA, Tucker-Heard G, Perdue NR, Killebrew JR, Urdahl KB, Campbell DJ (2009) The transcription factor T-bet controls regulatory T cell homeostasis and function during type 1 inflammation. Nat Immunol 10:595–602

Koyanagi M, Baguet A, Martens J, Margueron R, Jenuwein T, Bix M (2005) EZH2 and histone 3 trimethyl lysine 27 associated with Il4 and Il13 gene silencing in Th1 cells. J Biol Chem 280:31470–31477

Krawczyk CM, Shen H, Pearce EJ (2007) Functional plasticity in memory T helper cell responses. J Immunol 178:4080–4088

Langrish CL, Chen Y, Blumenschein WM, Mattson J, Basham B, Sedgwick JD, McClanahan T, Kastelein RA, Cua DJ (2005) IL-23 drives a pathogenic T cell population that induces autoimmune inflammation. J Exp Med 201:233–240

Lazarevic V, Chen X, Shim JH, Hwang ES, Jang E, Bolm AN, Oukka M, Kuchroo VK, Glimcher LH (2011) T-bet represses T(H)17 differentiation by preventing Runx1-mediated activation of the gene encoding RORgammat. Nat Immunol 12:96–104

Lee HJ, Takemoto N, Kurata H, Kamogawa Y, Miyatake S, O'Garra A, Arai N (2000) GATA-3 induces T helper cell type 2 (Th2) cytokine expression and chromatin remodeling in committed Th1 cells. J Exp Med 192:105–115

Lee GR, Fields PE, Flavell RA (2001) Regulation of IL-4 gene expression by distal regulatory elements and GATA-3 at the chromatin level. Immunity 14:447–459

Lee GR, Kim ST, Spilianakis CG, Fields PE, Flavell RA (2006) T helper cell differentiation: regulation by cis elements and epigenetics. Immunity 24:369–379

Lewis MD, Miller SA, Miazgowicz MM, Beima KM, Weinmann AS (2007) T-bet's ability to regulate individual target genes requires the conserved T-box domain to recruit histone methyltransferase activity and a separate family member-specific transactivation domain. Mol Cell Biol 27:8510–8521

Li B, Greene MI (2007) FOXP3 actively represses transcription by recruiting the HAT/HDAC complex. Cell Cycle 6:1432–1436

Lugo-Villarino G, Maldonado-Lopez R, Possemato R, Penaranda C, Glimcher LH (2003) T-bet is required for optimal production of IFN-gamma and antigen-specific T cell activation by dendritic cells. Proc Natl Acad Sci USA 100:7749–7754

Makar KW, Perez-Melgosa M, Shnyreva M, Weaver WM, Fitzpatrick DR, Wilson CB (2003) Active recruitment of DNA methyltransferases regulates interleukin 4 in thymocytes and T cells. Nat Immunol 4:1183–1190

Mikkelsen TS, Ku M, Jaffe DB, Issac B, Lieberman E, Giannoukos G, Alvarez P, Brockman W, Kim TK, Koche RP et al (2007) Genome-wide maps of chromatin state in pluripotent and lineage-committed cells. Nature 448:553–560

Miller SA, Weinmann AS (2010) Molecular mechanisms by which T-bet regulates T-helper cell commitment. Immunol Rev 238:233–246

Miller SA, Huang AC, Miazgowicz MM, Brassil MM, Weinmann AS (2008) Coordinated but physically separable interaction with H3K27-demethylase and H3K4-methyltransferase activities are required for T-box protein-mediated activation of developmental gene expression. Genes Dev 22:2980–2993

Miller SA, Mohn SE, Weinmann AS (2010) Jmjd3 and UTX play a demethylase-independent role in chromatin remodeling to regulate T-box family member-dependent gene expression. Mol Cell 40:594–605

Mosammaparast N, Shi Y (2010) Reversal of histone methylation: biochemical and molecular mechanisms of histone demethylases. Annu Rev Biochem 79:155–179

Mosmann TR, Cherwinski H, Bond MW, Giedlin MA, Coffman RL (1986) Two types of murine helper T cell clone. I. Definition according to profiles of lymphokine activities and secreted proteins. J Immunol 136:2348–2357

Mukasa R, Balasubramani A, Lee YK, Whitley SK, Weaver BT, Shibata Y, Crawford GE, Hatton RD, Weaver CT (2010) Epigenetic instability of cytokine and transcription factor gene loci underlies plasticity of the T helper 17 cell lineage. Immunity 32:616–627

Murphy KM, Stockinger B (2010) Effector T cell plasticity: flexibility in the face of changing circumstances. Nat Immunol 11:674–680

Murphy E, Shibuya K, Hosken N, Openshaw P, Maino V, Davis K, Murphy K, O'Garra A (1996) Reversibility of T helper 1 and 2 populations is lost after long-term stimulation. J Exp Med 183:901–913

Naiche LA, Harrelson Z, Kelly RG, Papaioannou VE (2005) T-box genes in vertebrate development. Annu Rev Genet 39:219–239

Nurieva RI, Chung Y, Martinez GJ, Yang XO, Tanaka S, Matskevitch TD, Wang YH, Dong C (2009) Bcl6 mediates the development of T follicular helper cells. Science 325:1001–1005

Oestreich KJ, Huang AC, Weinmann AS (2011). The lineage-defining factors T-bet and Bcl-6 collaborate to regulate Th1 gene expression patterns. J Exp Med 208:1001–1013

Oldenhove G, Bouladoux N, Wohlfert EA, Hall JA, Chou D, Dos Santos L, O'Brien S, Blank R, Lamb E, Natarajan S et al (2009) Decrease of Foxp3+ Treg cell number and acquisition of effector cell phenotype during lethal infection. Immunity 31:772–786

O'Shea JJ, Paul WE (2010) Mechanisms underlying lineage commitment and plasticity of helper CD4+ T cells. Science 327:1098–1102

Osorio F, LeibundGut-Landmann S, Lochner M, Lahl K, Sparwasser T, Eberl G, Reis e Sousa C (2008) DC activated via dectin-1 convert Treg into IL-17 producers. Eur J Immunol 38: 3274–3281

Pan F, Yu H, Dang EV, Barbi J, Pan X, Grosso JF, Jinasena D, Sharma SM, McCadden EM, Getnet D et al (2009) Eos mediates Foxp3-dependent gene silencing in CD4+ regulatory T cells. Science 325:1142–1146

Reinhardt RL, Kang SJ, Liang HE, Locksley RM (2006) T helper cell effector fates—who, how and where? Curr Opin Immunol 18:271–277

Ruthenburg AJ, Allis CD, Wysocka J (2007) Methylation of lysine 4 on histone H3: intricacy of writing and reading a single epigenetic mark. Mol Cell 25:15–30

Schoenborn JR, Wilson CB (2007) Regulation of interferon-gamma during innate and adaptive immune responses. Adv Immunol 96:41–101

Schoenborn JR, Dorschner MO, Sekimata M, Santer DM, Shnyreva M, Fitzpatrick DR, Stamatoyannopoulos JA, Wilson CB (2007) Comprehensive epigenetic profiling identifies multiple distal regulatory elements directing transcription of the gene encoding interferon-gamma. Nat Immunol 8:732–742

Shevach EM (2009) Mechanisms of foxp3+ T regulatory cell-mediated suppression. Immunity 30:636–645

Shi Y (2007) Histone lysine demethylases: emerging roles in development, physiology and disease. Nat Rev Genet 8:829–833

Shilatifard A (2006) Chromatin modifications by methylation and ubiquitination: implications in the regulation of gene expression. Annu Rev Biochem 75:243–269

Szabo SJ, Kim ST, Costa GL, Zhang X, Fathman CG, Glimcher LH (2000) A novel transcription factor, T-bet, directs Th1 lineage commitment. Cell 100:655–669

Szabo SJ, Sullivan BM, Stemmann C, Satoskar AR, Sleckman BP, Glimcher LH (2002) Distinct effects of T-bet in TH1 lineage commitment and IFN-gamma production in CD4 and CD8 T cells. Science 295:338–342

Vignali DA, Collison LW, Workman CJ (2008) How regulatory T cells work. Nat Rev Immunol 8:523–532

Wei G, Wei L, Zhu J, Zang C, Hu-Li J, Yao Z, Cui K, Kanno Y, Roh TY, Watford WT et al (2009) Global mapping of H3K4me3 and H3K27me3 reveals specificity and plasticity in lineage fate determination of differentiating CD4+ T cells. Immunity 30:155–167

Wilson CB, Rowell E, Sekimata M (2009) Epigenetic control of T-helper-cell differentiation. Nat Rev Immunol 9:91–105

Yen D, Cheung J, Scheerens H, Poulet F, McClanahan T, McKenzie B, Kleinschek MA, Owyang A, Mattson J, Blumenschein W et al (2006) IL-23 is essential for T cell-mediated colitis and promotes inflammation via IL-17 and IL-6. J Clin Invest 116:1310–1316

Yu D, Vinuesa CG (2010) The elusive identity of T follicular helper cells. Trends Immunol 31:377–383

Yu D, Rao S, Tsai LM, Lee SK, He Y, Sutcliffe EL, Srivastava M, Linterman M, Zheng L, Simpson N et al (2009) The transcriptional repressor Bcl-6 directs T follicular helper cell lineage commitment. Immunity 31:457–468

Zheng W, Flavell RA (1997) The transcription factor GATA-3 is necessary and sufficient for Th2 cytokine gene expression in CD4 T cells. Cell 89:587–596

Zheng Y, Josefowicz SZ, Kas A, Chu TT, Gavin MA, Rudensky AY (2007) Genome-wide analysis of Foxp3 target genes in developing and mature regulatory T cells. Nature 445:936–940

Zhou W, Chang S, Aune TM (2004) Long-range histone acetylation of the IFNg gene is an essential feature of T cell differentiation. Proc Natl Acad Sci USA 101:2440–2445

Zhou L, Lopes JE, Chong MM, Ivanov II, Min R, Victora GD, Shen Y, Du J, Rubtsov YP, Rudensky AY et al (2008) TGF-beta-induced Foxp3 inhibits T(H)17 cell differentiation by antagonizing RORgammat function. Nature 453:236–240

Zhou L, Chong MM, Littman DR (2009) Plasticity of CD4+ T cell lineage differentiation. Immunity 30:646–655

Zhu J, Paul WE (2008) CD4 T cells: fates, functions, and faults. Blood 112:1557–1569

Zhu J, Yamane H, Paul WE (2010) Differentiation of effector CD4 T cell populations. Annu Rev Immunol 28:445–489

The Epigenetic Landscape of Lineage Choice: Lessons From the Heritability of *Cd4* and *Cd8* Expression

Manolis Gialitakis, MacLean Sellars and Dan R. Littman

Abstract Developing $\alpha\beta$ T cells choose between the helper and cytotoxic lineages, depending upon the specificity of their T cell receptors for MHC molecules. The expression of the CD4 co-receptor on helper cells and the CD8 co-receptor on cytotoxic cells is intimately linked to this decision, and their regulation at the transcriptional level has been the subject of intense study to better understand lineage choice. Indeed, as the fate of developing T cells is decided, the expression status of these genes is accordingly locked. Genetic models have revealed important transcriptional elements and the ability to manipulate these elements in the framework of development has added a new perspective on the temporal nature of their function and the epigenetic maintenance of gene expression. We examine here novel insights into epigenetic mechanisms that have arisen through the study of these genes.

Manolis Gialitakis and MacLean Sellars contributed equally to this work

---

M. Gialitakis · M. Sellars · D. R. Littman (✉)
Molecular Pathogenesis Program,
Howard Hughes Medical Institute,
Kimmel Center for Biology and Medicine at the Skirball
Institute of Biomolecular Medicine,
New York University School of Medicine,
New York, NY 10016, USA
e-mail: dan.littman@med.nyu.edu

Current Topics in Microbiology and Immunology (2012) 356: 165–188
DOI: 10.1007/82_2011_175
© Springer-Verlag Berlin Heidelberg 2011
Published Online: 12 October 2011

Contents

1 Introduction 166
   1.1   The T cell Helper versus Cytotoxic Lineage Choice as a Model
       for Bi-potential Fate Decisions 166
   1.2   Molecular Mechanism of Transcriptional Regulation
       and Epigenetic Propagation 168
2 Epigenetic Regulation of Co-Receptor Loci 171
   2.1   Epigenetic Regulation of the *Cd4* Locus 171
   2.2   Epigenetic Regulation of the *Cd8* Locus 178
   2.3   Long Distance Interactions between *Cd4* and *Cd8* Co-Receptor
       Loci During Lineage Choice 181
3 Concluding Remarks 181
References 182

1 Introduction

1.1 The T cell Helper versus Cytotoxic Lineage Choice as a Model for Bi-potential Fate Decisions

Development of even the most complex organisms can be broken down into a series of bi-potential fate decisions: apoptosis versus survival, proliferation versus quiescence, differentiation versus renewal, etc. During differentiation, these choices yield cells that are increasingly restricted in lineage potential until a terminal, functional fate is reached (i.e. a neuron, an epithelial cell, a helper T cell, etc.). While some stages of differentiation are plastic, bifurcation points are reached at which a cell cannot reverse course and take on alternative fates. Mechanistically, this involves the activation of a lineage-specific transcriptional program (specification) and repression of the programs of alternative lineages (commitment). In many cases, these transcriptional programs must stably endure many rounds of mitosis. This is in part achieved through the binding of sequence-specific transcription factors, but is also thought to be regulated by heritable epigenetic marks, which overlay important lineage information on primary DNA sequence. As we discuss below, TCR$\alpha\beta$ T cell development, and specifically the choice between the CD4$^+$ helper and the CD8$^+$ cytotoxic T cell fates, is an ideal model for studying the epigenetic mechanisms of bi-potential decisions and their maintenance.

The majority of T cells in the body expresses TCR$\alpha\beta$ and develop in the thymus from bone marrow-derived precursors [reviewed in Rothenberg et al. (2008)]. At the earliest developmental stages, these T cell progenitors are referred to as double negatives (DN), owing to a lack of CD4 and CD8 co-receptor expression on the cell surface. DN cells proceed through multiple stages of lineage restriction and differentiation, including commitment to the TCR$\alpha\beta$, rather than the TCR$\gamma\delta$ lineage [reviewed in Ciofani and Zuniga-Pflucker 2010)]. At the DN stage, cells rearrange the gene encoding the TCR β chain, and in-frame productive VDJ rearrangement of one allele allows cells to pass the β selection checkpoint and proceed through multiple rounds of division. β-selected cells then up-regulate CD4 and CD8, becoming CD4$^+$CD8$^+$ double positive (DP) cells. DPs commence TCR α chain gene rearrangements and eventually follow one of three fates tied to the TCR: (1) unsuccessful *Tcra* rearrangement results in death by neglect; (2) rearrangements that yield TCRs with high avidity for self peptide-MHC result in negative selection; and (3) rearrangements that yield TCRs of intermediate avidity result in positive selection [reviewed in Starr et al. (2003)].

Following positive selection, T cells commit to either the CD4$^+$CD8$^-$ helper lineage or the CD8$^+$CD4$^-$ cytotoxic lineage, depending on MHC specificity. CD4 and CD8 are co-receptors for MHCII and MHCI, respectively, and during lineage choice their expression is matched to TCR specificity for either class of major histocompatibility molecules. How TCR specificity for MHC translates into co-receptor expression and lineage choice is still a matter of debate and multiple models have been proposed [reviewed in Singer et al. (2008)]. Early models suggested that the decision was either stochastic, or instructive through quantitatively or qualitatively different signals. It is now thought more likely that lineage choice is instructed by the duration of co-receptor facilitated TCR-MHC signaling. Following positive selection, all T cells downregulate *Cd8* transcription, becoming CD4$^+$CD8^{lo} cells and attenuating potential signaling through MHCI-specific TCRs (Sarafova et al. 2005). Importantly, placing *Cd4* expression under the control of *Cd8* regulatory elements results in MHCII-specific cytotoxic T cells (Sarafova et al. 2005). This result is consistent with a shorter signal, following positive selection of MHCI-specific TCRs, leading to the cytotoxic T cell fate, and a longer lasting signal, through recognition of MHCII, resulting in the helper T cell fate.

Significant progress has been made in recent years in understanding the transcriptional network that underlies CD4$^+$ versus CD8$^+$ lineage choice and commitment [reviewed in Naito and Taniuchi (2010)]. Three transcription factors have emerged as being especially important. Gata3 appears to specify the helper lineage, which is subsequently sealed by the action of the BTB-POZ transcription factor ThPOK. Gata3 upregulates ThPOK, but that activity is not sufficient, as it is additionally required for helper cell differentiation independently of ThPOK (Hernandez-Hoyos et al. 2003; Pai et al. 2003; Wang et al. 2008b). ThPOK is required for CD4 T cell development, and constitutive ThPOK expression that can redirect MHCI-specific T cells to the helper lineage in a GATA3-dependent manner (He et al. 2005; Sun et al. 2005; Wang et al. 2008b). It is thought that ThPOK functions as a commitment factor, antagonizing the CD8-specific transcriptional program (Egawa and Littman 2008; Muroi et al. 2008; Wang et al. 2008a). One of the factors that ThPOK antagonizes is the runt domain transcription factor Runx3, which is critical for cytotoxic lineage development (Egawa et al. 2007; Taniuchi et al. 2002a). Runx3, analogous to ThPOK, is thought to act as a commitment factor that suppresses the helper transcriptional program (Egawa and Littman 2008). In addition, Runx3 is required for reactivating *Cd8* expression and silencing *Cd4* expression in MHCI selected cells, as will be discussed in more detail later (Sato et al. 2005; Taniuchi et al. 2002a, b). In addition to these critical regulators, a number of other transcription factors have been reported to play a role in lineage choice, including Myb and Tox in helper cell development, and MAZR and STAT5 (downstream of IL-7 signaling) in cytotoxic T cell development [reviewed in Naito and Taniuchi (2010)].

Considering the exquisite correlation between CD4 and CD8 expression and helper and cytotoxic lineage commitment, respectively, studying the transcriptional regulation of *Cd4* and *Cd8* has long been a strategy to uncover the factors that define lineage choice. Indeed the importance of Runx and MAZR proteins to lineage choice was first identified through examination of the *Cd4* and *Cd8* loci, respectively (Bilic et al. 2006; Taniuchi et al. 2002a). The fact that the loci are transiently co-expressed in DP cells and stably expressed or repressed in helper and cytotoxic cells, also makes them an excellent model to study transcriptional regulation and the mechanisms that control temporary versus permanent gene expression states. Below, we will review recent advances in the epigenetic regulation of the *Cd4* and *Cd8* loci; these provide insights into general mechanisms of transcriptional regulation during differentiation.

1.2 Molecular Mechanism of Transcriptional Regulation and Epigenetic Propagation

To fit the few meters of DNA that exist in every eukaryotic nucleus, cells must tightly pack their genetic material. The nucleosome, consisting of 147bp of DNA wrapped around a histone octomer, is the basic packaging unit and forms the classic "beads on a string" structure observed in electron microscopy images (Olins and Olins 1974). Histone proteins H2A, H2B, H3 and H4 (two each in a histone octamer) are positively charged and have intrinsic DNA binding affinity, independent of nucleotide sequence. Histone H1 binds between adjacent nucleosomes, creating a higher order structure called the 30nm chromatin fiber [reviewed in Woodcock and Ghosh (2010)].

Nucleosomal packaging can physically impede transcription by RNA polymerase and thus different degrees of packing can modulate transcriptional outcomes [reviewed in Orphanides and Reinberg (2000)]. Looser packed, transcriptionally permissive chromatin is termed "euchromatin", while more tightly packed, repressive chromatin is called "heterochromatin". These chromatin states are dynamic during development and in response to extracellular signals. Covalent modifications of histone proteins (especially their protruding aminoterminal tails) and DNA are thought to be the biochemical basis for the distinction between heterochromatin and euchromatin. Initially, histones were found to be acetylated and this correlated with looser wrapping of the DNA (Allfrey 1966). Additional post-translational modifications have been identified, including methylation, phosphorylation, ubiquitination and sumoylation. Genome-wide study of such modifications has allowed the correlation of some specific modifications with gene activity (Barski et al. 2007; Heintzman et al. 2007; Ji et al. 2010; Roh et al. 2007). For example, tri-methylation of Histone H3 lysine 9 (H3K9), H3K27,

H4K20 and H3K79 are associated with repressed chromatin states and hetero-chromatin. In contrast, H3K4me1-3, histone acetylation (H3Ac and H4Ac), and mono-methylation of H3K9 and H3K27 are associated with gene activation and euchromatic regions. Thus, different modifications on different residues, the same modification on different residues and the abundance of a single modified residue may all have unique effects on transcription. In addition to histones, DNA itself can be modified by methylation on cytosine residues, a mark that has generally been correlated with gene silencing (Bachman et al. 2003; Fuks et al. 2000), but may also facilitate transcription in certain contexts (Wu et al. 2010).

Chromatin modifications do not act in a vacuum; their effect on transcription depends not only on the specific residue modified, but also on the surrounding residues and their modifications [reviewed in Campos and Reinberg (2009)]. Further, combinations of multiple modifications may result in a context-dependent outcome on transcription [reviewed in Lee et al. (2010)]. For example, H3K9me3 generally recruits HP1 to repress transcription, but in combination with phosphorylation of the adjacent H3S10, HP1 binding is abrogated and repression may be relieved (Fischle et al. 2005; Mateescu et al. 2008). Thus histone and DNA modifications superimpose a rich layer of information about the underlying DNA sequence.

Several enzyme classes, usually in the context of multi-factor complexes, catalyze chromatin modifications. Generally, these complexes do not recognize specific DNA sequences, but are recruited by sequence-specific transcription factors or recognition of specific histone modifications. Further, complexes exist to write and erase most chromatin modifications. Histone acetyltransferases (HAT) add acetyl groups, while histone deacetylase (HDAC) enzymes eliminate them. Enzymes also exist to methylate and demethylate lysines and arginines. Even DNA methylation, which is considered the most stable modification, may be erased by the recently characterized Tet proteins which can modify 5-methylcyt-osine, resulting in a 5-hydroxymethylcytosine that may prevent maintenance DNA methylation, or may be replaced by an unmethylated cytosine residue through a base excision repair pathway (Tahiliani et al. 2009); alternatively, 5-hydroxym-ethylcytosine could have a stand-alone function (Ficz et al. 2011). Taken together, this means that chromatin modifications can be dynamically written and erased from the genome in the process of regulating transcription.

The functional outcome of DNA and histone modifications is at least in part brought about through their interaction with specific protein motifs. Acetylated lysines are recognized by bromodomains, while chromodomains, PHD fingers, and WD40 repeats bind methylated lysines. Meanwhile, methylated cytosines are read by methyl binding domains. Engagement of these modifications by individual protein subunits may recruit or modulate the activity of multi-factorial chromatin modifying complexes. Importantly, several proteins can read different modifica-tions simultaneously to regulate their binding to nucleosomes (Bartke et al. 2010). Thus the "histone language" can be read with the aid of multiple adaptors, linking histone, and DNA marks to functional outcome.

Transcriptional activity, in addition to being correlated with specific histone modifications, has been correlated with nucleosome depletion at promoter regions

(Lee et al. 2004; Ozsolak et al. 2007) prior to transcriptional initiation (Petesch and Lis 2008). It has long been known that functional DNA elements are hypersensitive to endonucleases, and such DNase I hypersensitive sites (DHS) (Wu et al. 1979) occur upon nucleosome depletion. Studies of *IFNβ* gene activation during viral infection revealed that a nucleosome masking the transcription start site (TSS) and TATA box was remodeled by the histone acetylation-recruited SWI/SNF chromatin remodeling complex, allowing for TFIID recruitment (Agalioti et al. 2000). Binding of the TFIID subunit TBP to the TATA box induced the nucleosome to slide further downstream, exposing the transcription start site and allowing for transcription (Lomvardas and Thanos 2001). Thus there is a complex interplay between chromatin marks, their readers, writers, and erasers, and sequence-specific transcription factors in the regulation of transcription.

Views of transcriptional regulation have changed dramatically in recent years. Transcription is no longer thought to occur only in a linear fashion with regulatory elements controlling the expression of their downstream genes. Rather, it is thought to be a dynamic process involving the movement of chromatin fibers to allow co-regulated genes to come together in subnuclear space. Osborne et al. (2004) found that active genes co-localize with reservoirs of active RNA polymerase II in what are termed "transcription factories". Subsequently, Spilianakis et al. showed that genes on different chromosomes (the *Il4* locus on chromosome 11 and *Ifng* on chromosome 10) could interact in a developmentally regulated manner and this interaction had a functional role in the expression of both loci, since deletion of a regulatory element on one chromosome could affect the expression of a locus on the other (2005). Such long-range movements can affect not only gene activity, but can also have an impact on the transcriptional competence of loci. For example, transient IFNγ signaling induces persistent association of the MHCII locus with promyelocytic leukemia (PML) nuclear bodies, which perpetuates histone marks through mitoses. Thus the locus is maintained in a poised state, sensitizing it to respond to lower doses of IFNγ, with faster kinetics (Gialitakis et al. 2010).

Where does epigenetic regulation fit into all of this? Differentiation is a dynamic process that relies on the inheritance of gene expression patterns. Although almost all cells within an organism have the same genetic material (antigen receptor loci excluded), the transcriptional outcomes, and thus cell type/lineage, can differ dramatically. An extra layer of stably inherited information, unique to each cell type, is provided by epigenetic modifications. Strictly speaking, epigenetic marks are not contained in the primary genetic sequence, but are stable and heritable even in the absence of their sequence-specific initiating events. For example, H3K27 trimethylation, is both catalyzed and recognized by the PRC2 methyltransferase complex (Hansen et al. 2008). The PRC2 subunit, EED, binds H3K27me3 both to recruit the PRC2 complex and to allosterically activate its methyltransferase activity, setting up a positive feedback loop to propagate the epigenetic mark through cell division without need for sequence-specific factors (Margueron et al. 2009). In this context, many chromatin modifications are not in fact epigenetic. They may be by-products of the current transcriptional state of a locus, or they may require the continued action of sequence-specific factors or events that initiated them. These are interesting to study in the context of how a gene is acutely transcribed or repressed, but may not help to explain how the transcriptional program of a helper or cytotoxic T cell can be stably maintained through many mitoses. In what follows, we discuss what the *Cd4* and *Cd8* loci have to teach us about heritable epigenetic regulation of gene expression.

2 Epigenetic Regulation of Co-Receptor Loci

2.1 Epigenetic Regulation of the Cd4 Locus

Cd4 expression is controlled by multiple regulatory elements (Fig. 1a). DHS mapping revealed several putative regulatory elements in the locus (Adlam and Siu 2003; Sands and Nikolic-Zugic 1992; Sawada and Littman 1991), and eventually led to the identification of three enhancers: the distal ($E4_D$) and proximal enhancers ($E4_P$) at ~ 24 and ~ 13 kb upstream of the *Cd4* TSS, respectively, and the thymocyte enhancer ($E4_T$) at ~ 36 kb downstream (Adlam and Siu 2003; Sawada and Littman 1991; Wurster et al. 1994) (Reviewed in detail by Taniuchi et al. (2004)]. In combination with the *Cd4* promoter, $E4_P$ drives T cell-specific transgene expression in multiple mouse lines (Blum et al. 1993; Hanna et al. 1994; Killeen et al. 1993). The developmental window during which $E4_P$ shows activity, however, was not entirely clear: while some studies found it to be active in DN and pre-selection DP cells through to mature helper and cytotoxic T cells (Blum et al. 1993; Hanna et al. 1994; Killeen et al. 1993; Manjunath et al. 1999; Sawada et al. 1994; Siu et al. 1994), others indicated that $E4_P$ activity begins post-positive selection (Adlam et al. 1997). In addition, it appears that, in the context of transgenes, $E4_P$ may lose activity in mature T cells following TCR stimulation (Manjunath et al. 1999). $E4_D$ exhibits T cell line-specific activity in transient transfection assays (Wurster et al. 1994), but its in vivo relevance for *Cd4* regulation has not been demonstrated and, moreover, the homologous human sequence has been implicated in control of the adjacent *LAG-3* gene (Bruniquel et al. 1998) and the combination of $E4_D$ and $E4_P$ in reporter transgenes drives expression in B cells and macrophages (Siu et al. 1994). Finally $E4_T$ has been suggested to drive expression in DPs, but only in combination with $E4_P$ (Adlam and Siu 2003). Taken together, these sometimes-conflicting studies have identified three possible *Cd4* enhancers and ascribed them independent, overlapping and cooperative functions at different developmental stages.

Germline and conditional targeting of two of these enhancers, $E4_P$ and $E4_T$, has helped to determine their relevant functions (Chong et al. 2010) (Fig. 1c). $E4_T$ was found to be dispensable for *Cd4* expression in TCR$\alpha\beta$ T cells, but required for expression on a subset of lymphoid tissue inducer (LTi) cells in the small intestine lamina propria, now commonly referred to as innate lymphoid cells. In contrast, $E4_P$ was required for *Cd4* expression in pre-selection DP cells. However, CD4 was expressed following positive selection in $E4_P$ thymocytes, suggesting the existence of a yet unidentified "maturation enhancer". Importantly, $E4_P$ activity was dispensable in the periphery, as its Cre-mediated deletion in mature CD4$^+$ T cells did

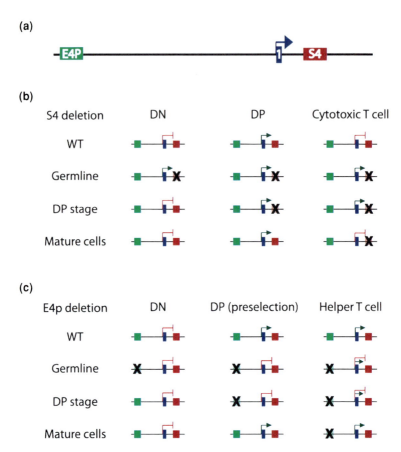

Fig. 1 Epigenetic regulation of *Cd4* transcription and silencing. **a** Genomic organization of the *Cd4* locus: proximal enhancer (E4$_P$, *green box*), exon 1 (1, *blue box*) and silencer (S4, *red box*). **b** S4 deletion at different developmental stages reveals epigenetic silencing of *Cd4*. In WT cells, *Cd4* is silenced in immature DN and mature cytotoxic T cells, but is expressed in DP cells. Germline deletion of S4 leads to inappropriate expression of *Cd4* in DN and cytotoxic T cells. Inducible S4 deletion at the transition to the DP stage results in expression in cytotoxic T cells. However, inducible S4 deletion in cytotoxic T cells does not lead to ectopic expression, indicating that *Cd4* is heritably silenced in mature cells (i.e. independently of S4). **c** E4$_P$ deletion at different developmental stages reveals epigenetic maintenance of *Cd4* expression. In WT mice, *Cd4* is expressed in DP cells and in mature helper T cells. E4$_P$ deletion in the germline, or at the transition between the DN and DP stages, results in DP cells that fail to express *Cd4* prior to positive selection, and in unstable *Cd4* expression in helper T cells. Inducible E4$_P$ deletion in helper cells does not affect *Cd4* expression, indicating epigenetic maintenance of expression

not affect *Cd4* expression. Moreover, E4$_P$ is required at the DP stage for stable, high-level CD4 expression in mature cells, as will be discussed later. Thus this study indicated that *Cd4* transcription is potentiated at the DP stage by E4$_P$, and suggested that, after positive selection, it is regulated by an unidentified maturation enhancer whose heritable activity would be initiated in concert with E4$_P$.

The Epigenetic Landscape of Lineage Choice

Restriction of *Cd4* expression to DPs and mature helper T cells is conferred by the activity of Runt domain-containing transcription factors that bind to sites in a silencer element (S4). S4 was initially identified as a 434 bp element in the first intron of *Cd4*, which suppressed transgene expression in DN and CD8$^+$ T cells (Sawada et al. 1994). Germline deletion of S4 resulted in expression of CD4 in all T cells starting at the DN stage, indicating that this element was responsible for suppressing developmental stage-inappropriate *Cd4* expression (Fig. 1b) (Leung et al. 2001; Zou et al. 2001). Subsequently, silencer activity was found to be mediated by Runx1 and 3, in DN and cytotoxic T cells, respectively (Taniuchi et al. 2002a, b). Germline mutations in Runx binding motifs in S4 led to CD4 expression in DN and mature CD8 thymocytes. Further, deletion of Runx1 revealed that it is indispensable for *Cd4* silencing in DN cells. In contrast, Runx3 deletion led to variegated *Cd4* expression in mature CD8$^+$ cells, indicating a role for Runx3-mediated silencing later in development. Thus Runx1 and 3 have stage-specific roles in mediating *Cd4* repression through S4.

In collaboration with other factors, Runx1 and 3 mediate two developmental stage-specific modes of *Cd4* silencing: reversible and permanent. In DN cells, silencing is reversible, as *Cd4* transcription must be activated upon transition to the DP stage. What is the mechanism? One model that can be constructed from recent data would involve active antagonism of E4$_P$ function by Runx1. The HEB and E2A bHLH transcription factors, which are crucial for E4$_P$-mediated *Cd4* activation between the DN and DP stages (Jones and Zhuang 2007; Sawada and Littman 1993), are preloaded onto E4$_P$ at the DN stage (Yu et al. 2008). In the presence of S4, however, p300 recruitment to E4$_P$, and thus transcriptional activation, is impaired (Yu et al. 2008). This antagonism of p300 recruitment could be mediated by long-range interactions between S4, bound by Runx1, and E4$_P$, as was recently reported (Jiang and Peterlin 2008). In accord with this model, the bHLH-ZIP transcription factor AP4 binds E4$_P$, interacts with Runx1, and is required for efficient silencing of *Cd4* in DN cells (Egawa and Littman 2011). Thus, a physical interaction between Runx1 and AP4 may bring E4$_P$ and S4 into close proximity, so that Runx1 or other silencer-associated factors can antagonize co-activator recruitment by HEB, E2A, and other positively acting factors. Additionally Runx1-AP4 mediated interaction between S4 and E4$_P$ could prevent recruitment of E4$_P$-bound transcriptional co-factors such as P-TEFb to the promoter, precluding elongation by RNA PolII until the DP stage (Jiang et al. 2005) (Fig. 2).

It should be noted that reversible *Cd4* silencing in DNs is likely more complicated than this model suggests. Genetic studies have demonstrated that the BAF57 and BRG subunits of the SWI/SNF-like chromatin remodeling complexes are critical for *Cd4* silencing in DN cells (Chi et al. 2002, 2003). Interestingly, dominant negative BAF57 expression or T cell-specific *Brg* deletion results in decreased chromatin accessibility and Runx1 binding at S4 accompanied by CD4 de-repression in DN cells (Wan et al. 2009), indicating that SWI/SNF contributes to reversible silencing by remodeling chromatin to allow for Runx1 recruitment. In contrast to SWI/SNF, the NuRD chromatin remodeling complex has been implicated in reversing *Cd4* silencing. Deletion of the Mi2b NuRD subunit allows

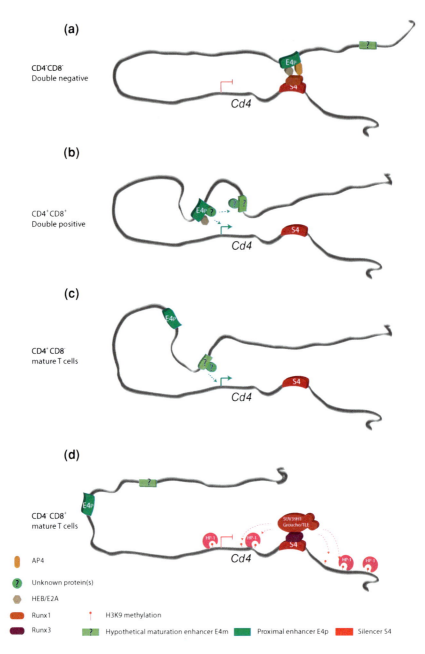

Cd4 silencing to continue past the DN stage (Naito et al. 2007). Taking into account that NuRD is generally considered a repressive chromatin remodeling complex, it is tempting to speculate that NuRD could remodel S4 chromatin to a state inaccessible to Runx1, eliminating silencer function during the transition from DN to DP. Taken together, it appears that reversible silencing requires

The Epigenetic Landscape of Lineage Choice

◀ **Fig. 2** A model for *Cd4* regulation during development. **a** At the DN stage, S4 interacts with the E4$_P$ enhancer, preventing it from interacting with, and activating, the *Cd4* promoter. This interaction could be mediated through association between E4$_P$-bound AP-4 and S4-bound Runx1. **b** In DP cells, S4 is inactivated, allowing E4$_P$ to interact with the *Cd4* promoter to drive transcription. Deletion of E4$_P$ suggests that it may activate a yet unidentified maturation enhancer (E4$_M$), or collaborate with this enhancer to activate an epigenetic state required for subsequent high-level, stable *Cd4* expression in mature helper T cells. **c** In mature helper cells, *Cd4* is expressed independently of E4$_P$, possibly due to epigenetic mechanisms or E4$_P$-mediated activation of the putative E4$_M$ enhancer. **d** In cytotoxic T cells, *Cd4* is silenced by S4 in a Runx3 and possibly HP-1 dependent manner. Runx3 could recruit transcriptional co-repressors such as SUV39H1 and Groucho/TLE to the locus to deposit repressive histone marks, such as Histone H3K9 methylation. These marks could in turn recruit HP-1 to epigenetically suppress transcription. The role of DNA methylation remains to be further investigated and thus is not shown here. TCF-1α/LEF-1 binding on E4$_P$ is not shown

chromatin remodeling by SWI/SNF, which allows Runx1 to bind to S4 and interact with E4$_P$-bound AP4, thus actively repressing transcriptional elongation at *Cd4*.

In contrast to transient *Cd4* silencing in DN cells, silencing in mature cytotoxic T cells appears to be permanent and mediated epigenetically (Fig. 1b). Mutation of individual critical transcription factor binding sites in S4 (including one Runx binding motif) led to partial, but uniform, CD4 de-repression in DN cells, but variegated CD4 expression on CD8$^+$ T cells (Taniuchi et al. 2002a, b). This variegated pattern is reminiscent of position effect variegation (PEV) of transgene expression, which is mediated by heterochromatin spreading from adjacent loci and its stable propagation through multiple mitoses (Fodor et al. 2010). The epigenetic nature of *Cd4* silencing was confirmed by the finding of continued stable silencing of CD4 expression through multiple rounds of cell division following Cre-mediated deletion of S4 in mature CD8$^+$ T cells (Zou et al. 2001). Thus S4 initiates *Cd4* silencing during development, and this silenced state may be epigenetically propagated in the absence of S4.

These data indicate that epigenetic silencing of *Cd4* involves two distinct mechanisms: (1) a silenced state is first initiated after positive selection by factors associated with S4, and (2) silencing is maintained in mature cells independent of S4. Runx3 binding to S4 is clearly required for the initiation of silencing, but not maintenance (Taniuchi et al. 2002a, b). To understand maintenance, we initially assessed DNA methylation, as it is required to maintain heterochromatin-dependent X chromosome inactivation (Sado et al. 2000). We found that pharmacological inhibition of DNA methyltransferase activity with 5-azacytidine did not induce CD4 expression in proliferating CD8$^+$ cells, and argued that DNA methylation does not have a key role in maintenance (Zou et al. 2001). However, the time frame of the experiment, with a relatively small number of mitoses, precluded reaching a definitive conclusion, and this issue needs to be further explored with mice mutant for DNA methyltransferase genes. Intriguingly, overexpression of the heterochromatin protein HP-1β partially rescued silencing in mice in which silencer function was compromised by deletion of the binding motif for a critical, but yet unidentified, transcription factor, indicating that HP-1 proteins, and the H3K9 methylation marks that they recognize, may contribute to silencing (Taniuchi et al. 2002b) (Fig. 2d). In accord with this finding, Runx transcription factors can associate with the Groucho/TLE corepressor complex and the SUV39H1 H3K9 methyltransferase to mediate repression (Levanon et al. 1998; Reed-Inderbitzin et al. 2006). The Runx3 VWRPY motif is required for interactions with Groucho/TLE (its importance for interactions with SUV39H1 is unknown) and for epigenetic $Cd4$ silencing in mature CD8$^+$ cells (Yarmus et al. 2006), indicating that the Groucho/TLE co-repressor or other factors that interact with Runx3 through this domain are required for silencing. Importantly, the Runx1 VWRPY motif is dispensable for its silencing function in DNs (Telfer et al. 2004), highlighting the two modes of $Cd4$ silencing: transient in DNs and permanent/epigenetic in CD8$^+$ T cells.

What other mechanisms could be involved in the initiation and maintenance of epigenetic $Cd4$ silencing? X chromosome inactivation (XCI) is reminiscent of this process as an X chromosome inactivation center (Xic) is crucial for XCI in the inner cell mass (in mice), but is not necessary to maintain the inactive X in a silenced state in more differentiated cells (Brown and Willard 1994; Wutz and Jaenisch 2000). Interestingly, the non-coding XIST (nc)RNA encoded within the Xic is required for initiation but not maintenance of XCI (Csankovszki et al. 1999). Similarly, imprinted genes are silenced on one parental allele through the action of ncRNAs (Reviewed in O'Neill 2005). Thus it will be interesting to determine whether ncRNAs play a role in either the initiation or maintenance of the silenced state of $Cd4$. An interesting possibility is that Runx3 or another silencer binding factor functions to tether an ncRNA to the $Cd4$ locus to initiate silencing, similar to the recently described role of YY1 in tethering XIST to the X chromosome undergoing inactivation (Jeon and Lee 2011).

As illustrated in this discussion, $Cd4$ silencing deserves vigorous study in the future, as it can provide potentially novel insight into the mechanisms of establishment and inheritance of gene repression. Importantly, this is a unique system to study these events as in contrast to XCI, which occurs stochastically on either of the two X chromosomes, $Cd4$ silencing occurs in a developmentally regulated manner dependent on extracellular signals (i.e. TCR specificity for MHCI). This dependency on extracellular cues may provide unique insights into the initiation of epigenetic silencing that may not be revealed by XCI initiation due to its stochastic nature. Finally, the $Cd4$ system is less difficult to work with: knowledge of the signals that induce $Cd4$ silencing, as well as the fact that T cell differentiation occurs in post-natal mice, render this system more easily manipulated than the embryonic inner cell mass where XCI is initiated.

In addition to silencing, recent work from our laboratory has shown that the active transcription state of $Cd4$ in helper T cells is also propagated epigenetically, i.e. independently of the genetic element that initially activates $Cd4$ transcription (Fig. 1c). As mentioned earlier, E4$_P$ is required for CD4 expression in DP thymocytes (Chong et al. 2010). However, positive selection of E4$_P^{-/-}$ thymocytes activates CD4 expression in T helper lineage cells, possibly through a putative "maturation enhancer", leading to a reduced population of CD4$^+$ helper T cells. These cells had a broader distribution and lower amount of CD4 expression than

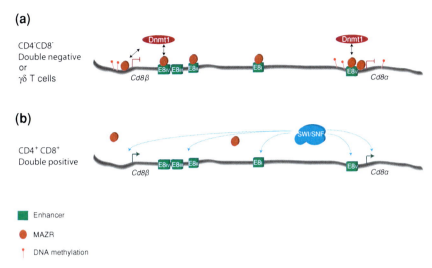

Fig. 3 A model for the epigenetic repression of *Cd8* locus transcription. **a** CpGs at the *Cd8a* and *Cd8b* promoters, as well as the E8$_V$ enhancer, are hypermethylated in DN and γδ T cells. Loss of this methylation due to DNMT1 deletion leads to inappropriate CD8 expression in these cells. MAZR associates with multiple regulatory elements in the *Cd8* locus, and represses *Cd8α/β* transcription at the DN to DP transition, possibly through interactions with DNA methyltransferases. **b** Downregulation of MAZR expression correlates with the relief of epigenetic *Cd8α/β* repression at the DP stage. Components of the SWI/SNF chromatin remodeling complex, and multiple enhancer elements (E8$_I$, E8$_{II}$, E8$_{III}$, and E8$_V$), contribute to the efficient activation of *Cd8α/β* expression in DP cells

wild-type cells, and lost expression upon TCR stimulation and proliferation. However, conditional deletion of E4$_P$ in mature peripheral CD4$^+$ cells by retroviral transduction of Cre had no effect on the stability or level of CD4 expression, even after many rounds of division. Thus, like the mirror opposite of the silencer, E4$_P$ sets an active epigenetic state of the *Cd4* locus in helper T cells, which can then be propagated in its absence.

The mechanisms underlying this positive epigenetic state are not entirely clear. Mature E4$_P^{-/-}$ CD4$^+$ T cells expressed lower levels of CD4 than WT despite normal levels of histone acetylation across the *Cd4* locus. In contrast, decreased *Cd4* transcription in these cells correlated with reduced H3K4me3 at the promoter. Upon TCR-induced proliferation, both histone modifications were lost in those E4$_P^{-/-}$ cells that lost *Cd4* expression. Considering that there was no concomitant increase in repressive histone marks such as H3K9me3 and H3K27me3 across the locus when compared to WT CD4$^+$ cells, it appears that instability of *Cd4* expression is due to the loss of activating marks rather than active silencing. Could differences in H3K4 methylation account for stable (WT) versus unstable (E4$_P^{-/-}$) memory? In yeast, H3K4me3 has been shown to persist after transcription has ceased, and thus has been postulated to serve as memory of previous transcriptional activity, though admittedly not through a cell cycle (Ng et al. 2003). Intriguingly, studies in mammalian cells have suggested longer memory through H3K4me3. Memory in the murine inflammatory response has been linked to persistent H3K4 trimethylation and H4 acetylation, and competence to recruit the SWI/SNF subunit Brg1 (Foster et al. 2007). In studies of IFNγ priming of MHCII gene transcription, H3K4me2 is stably transmitted through cell cycles, correlating with increased responsiveness to secondary stimulation (Gialitakis et al. 2010). Thus, one simple model would be that H3K4 methylation might allow transmission of the active state through the cell cycle, though the exact mechanism is still unclear.

What are other possible mechanisms for the propagation of active *Cd4* transcription? Two obvious suspects come to mind: DNA hypomethylation and deposition of histone variants. DNA demethylation is involved in activating epigenetic memory in the *Il2* and *FoxP3* loci. Activation of naïve T cells leads to demethylation of the *Il2* promoter, which in turn allows for a faster and stronger transcriptional response following secondary stimulation (Bruniquel and Schwartz 2003; Murayama et al. 2006). At the *FoxP3* locus, DNA demethylation allows binding of positively acting factors including Runx1 and FoxP3 itself, completing a feed-forward loop, which stabilizes FoxP3 expression (Bruno et al. 2009; Lal et al. 2009; Polansky et al. 2008; Williams and Rudensky 2007; Zheng et al. 2010). In the case of *Cd4* expression, E4$_P$-dependent hypomethylation of specific motifs in the locus may allow a transcription factor(s) to bind and maintain *Cd4* expression. Another, non-mutually exclusive, possibility is that deposition of specific histone variants potentiates memory. The variant H2A.Z has been shown to be required for memory (faster reactivation) of genes transcription in yeast (Brickner et al. 2007). While no difference in H2A.Z occupancy was found at the *Cd4* promoter between WT and E4$_P^{-/-}$ CD4$^+$ cells (Chong et al. 2010), it remains possible that H2A.Z underlies stable *Cd4* expression as it is also thought to play critical roles in enhancer accessibility (He et al. 2010). Thus E4$_P$-dependent H2A.Z deposition at an unidentified enhancer element could promote stable *Cd4* expression. Another candidate is the histone variant H3.3, which acts at the MyoD promoter to allow persistent (and inappropriate) MyoD expression in Xenopus muscle cell nuclei that have been reprogramed by two rounds of nuclear transfer into embryos (Ng and Gurdon 2008). Finer examination of DNA methylation and histone variant occupancy in WT and E4$_P^{-/-}$ cells will be required to evaluate these possibilities.

The mechanisms underpinning epigenetic memory of *Cd4* transcription are still unclear. Nevertheless, this system clearly deserves to be studied extensively, since, despite the widespread notion that self-propagating epigenetic mechanisms are critical to the developmental regulation of many genes, the activation and silencing of *Cd4* remain rare examples of *bona fide* heritability in mammalian gene expression.

2.2 Epigenetic Regulation of the Cd8 Locus

The *Cd8* locus consists of the *Cd8a* and *Cd8b* genes, separated by ~35kb in mice, and ~25kb in humans. CD8 is expressed either as a homo-dimer of CD8αα molecules, for example on intraepithelial lymphocytes (IEL) and CD8$^+$ DCs, or as

The Epigenetic Landscape of Lineage Choice

a heterodimer of CD8$\alpha\beta$ molecules on DP thymocytes and TCR$\alpha\beta$ cytotoxic T cells. Thus *Cd8a* and *Cd8b* genes can be both co-regulated and independently regulated [reviewed in Taniuchi et al. (2004)]. Here we focus mainly on *Cd8* locus regulation in the TCR$\alpha\beta$ lineage, and what hints this gives us into epigenetic mechanisms of gene regulation.

Cd8 locus expression is controlled by multiple enhancers. While the *Cd8a* promoter was not sufficient to drive lineage-specific expression, an 80kb fragment of the mouse locus stretching from 2kb upstream of *Cd8b* to 25kb downstream of *Cd8a* could drive tissue-specific expression (Hostert et al. 1997a). Thus all cis-regulatory elements critical for appropriate *Cd8a/b* expression were present in this interval. This study also identified four DHS clusters (CI-IV), as putative regulatory regions. Further transgenic studies based on these DHS clusters identified specific regions that regulate *Cd8* expression (Ellmeier et al. 1997, 1998; Hostert et al. 1998, 1997a, b; Kieffer et al. 1996, 1997; Zhang et al. 1998, 2001). The E8$_I$ (CIII-1, 2) enhancer drove transgene expression in mature CD8$^+$ cells and in IEL (Ellmeier et al. 1997, 1998; Hostert et al. 1997b). The E8$_{III}$ (CIV-3) enhancer drove expression in DP thymocytes, and E8$_{II}$ (CIV-4,5) in DPs and mature CD8$^+$ cells. The E8$_{IV}$ (CIV-1,2) element was more promiscuous than the above enhancers, driving low-level expression in CD4$^+$ T cells as well as DP and CD8$^+$ T cells (Ellmeier et al. 1998; Feik et al. 2005). Finally, while E8$_V$ (CII) exhibited no enhancer function by itself, a combination of E8$_V$ and E8$_I$ drove expression in DP cells in addition to mature CD8$^+$ cells (Hostert et al. 1998, 1997b). These studies indicated that regulation of *Cd8* locus expression in a cell type- and stage-specific manner is achieved through complex interactions between multiple and sometimes apparently redundant regulatory elements.

To determine the in vivo function of the individual *Cd8* locus enhancers in controlling *Cd8a* and *Cd8b* expression, knockout studies were undertaken. Deletion of E8$_I$, which drives transgene activity in IELs and mature CD8$^+$ cells, reduced CD8$\alpha\alpha$ and CD8$\alpha\beta$ expression on IEL by 40–80%, but left CD8 expression on TCR$\alpha\beta$ DPs and mature CD8$^+$ cells largely unaffected; there was only a minor decrease (10–20%) in CD8 expression at the CD8SP stage (Ellmeier et al. 1998; Hostert et al. 1998). There was no detectable phenotype in E8$_{II}^{-/-}$ animals, suggesting that loss of E8$_{II}$ activity may be compensated for by other enhancers (Ellmeier et al. 2002). Indeed, combined deletion of E8$_{II}$ and E8$_I$, or E8$_{II}$ and E8$_{III}$, resulted in variegated CD8 expression in DP cells (Ellmeier et al. 2002; Feik et al. 2005), indicating incomplete relief of heterochromatin-mediated *Cd8* repression. Combined E8$_{II}$/E8$_I$ deletion also reduced CD8 expression by 30% on mature CD8$^+$ T cells in the periphery. Similar to the combined deletions above, elimination of E8$_V$ resulted in variegated CD8 expression on DPs, as well as reduced CD8 expression on mature CD8$^+$ cells (down 20%) (Garefalaki et al. 2002). These results indicate that while E8$_I$ is specifically required for CD8$\alpha\alpha$ expression on IEL and DCs, the E8$_I$, E8$_{II}$, E8$_{III}$, and E8$_V$ elements all contribute to relieving heterochromatin-mediated *Cd8* locus repression at the DN to DPs stage, and to maintaining high-level CD8$\alpha\beta$ expression in mature CD8$^+$ cells (Fig. 3).

This begs the question: why does the *Cd8* locus contain multiple, seemingly redundant enhancers? Recent work in *Drosophila* demonstrated that apparently redundant enhancers are critical to maintaining expression in response to genetic and environmental stimuli (Frankel et al. 2010). Considering the importance of precisely controlled CD8 expression in lineage choice and cytotoxic T cell development (Fung-Leung et al. 1991; Sarafova et al. 2005), it is possible that this may also be the evolutionary driving force behind the development of multiple *Cd8* locus enhancers. Another possibility is that seemingly overlapping *Cd8* locus enhancers are important for the expression of *Cd8a* or *Cd8b* individually (Taniuchi et al. 2004). Most of the above work relied on surface CD8$\alpha\beta$ protein detection to infer *Cd8* gene expression, but murine CD8β cannot be expressed on the surface in the absence of CD8α (Devine et al. 2000). Thus, examination of *Cd8a* and *Cd8b* transcription in various enhancer knockout mouse strains would be required to evaluate whether individual enhancers are important for the expression of either gene individually.

CD8 expression appears to be regulated through multiple epigenetic mechanisms. The variegated CD8 expression phenotypes of DP thymocytes from E8$_{I/II}$, E8$_{II/III}$ and CII knockout mice suggest that the *Cd8* locus becomes activated during the DN to DP transition through reversal of potentially heritable repressive marks (Ellmeier et al. 2002; Feik et al. 2005; Garefalaki et al. 2002). Interestingly, expression of a dominant negative mutant of the Baf57 SWI/SNF complex subunit, or haploinsufficiency of the Brg subunit, resulted in diminished CD8 expression on DPs (Chi et al. 2002). Further, combining these two genetic defects in SWI/SNF members resulted in variegated CD8 expression on DPs. Thus, the SWI/SNF complex is critical for activation of CD8 expression at the transition from the DN4 to the DP stage (Fig. 3).

The activation of *Cd8* expression is not only controlled by positive regulators; DNA methylation and the MAZR zinc finger protein (*zfp278*) also play critical roles in epigenetically repressing the locus (Fig. 3). MAZR is highly expressed in DN cells, and is downregulated at the DP stage as CD8 expression is activated (Bilic et al. 2006). Forced expression of MAZR results in variegated CD8 expression at the DP stage (Bilic et al. 2006). Further, variegated CD8 expression on E8$_{I/II}$ deficient DP cells (Ellmeier et al. 2002) was partially relieved by MAZR deletion (Sakaguchi et al. 2010), suggesting that E8$_{I/II}$ antagonizes MAZR-mediated *Cd8* silencing during the DN to DP transition. This effect may be mediated directly, as MAZR binding has been observed at multiple elements in the *Cd8* locus (Bilic et al. 2006). Analysis of CD8 expression in mice with combined deletion of MAZR and various *Cd8* locus elements could reveal elements through which MAZR functions to repress *Cd8*. Interestingly, deletion of the maintenance DNA methyltransferase Dnmt1 on an E8$_{I/II}$ deficient background also partially rescued variegated CD8 expression (Bilic et al. 2006). While it is tempting to speculate that MAZR contributes to *Cd8* repression through recruitment or maintenance of DNA methylation, no clear link has been established between MAZR and initiating or maintenance DNA methyltransferases. In keeping with the observation that Dnmt1 contributes to *Cd8* locus repression, E8$_V$ sequences are differentially methylated between CD8$^+$ and CD8$^-$ cells (e.g. WT DP cells vs. liver or CD8$^-$ E8$_{I/II}$ DP cells) (Bilic et al. 2006; Carbone et al. 1988; Hamerman et al. 1997; Lee et al. 2001).

Further, *Dnmt1* deletion results in ectopic CD8 expression on TCR$\gamma\delta$ cells (Lee et al. 2001). Thus, DNA methylation silences *Cd8a/b* expression outside of the TCR$\alpha\beta$ lineage, and, along with MAZR, maintains *Cd8a/b* repression until the DP stage, apparently through epigenetic mechanisms.

2.3 Long Distance Interactions between Cd4 and Cd8 Co-Receptor Loci During Lineage Choice

In addition to *cis*-regulatory elements in each locus, it appears that *Cd4* and *Cd8* expression may be regulated by long-distance interactions between the loci and specific nuclear compartments. For example, the majority of *Cd4* alleles associate with peri-centromeric heterochromatin (PCH) in CD8$^+$ but not CD4$^+$ T cells; similarly *cd8* alleles preferentially associate with PCH in CD4$^+$ but not CD8$^+$ T cells (Collins et al. 2011; Delaire et al. 2004; Merkenschlager et al. 2004). These results indicate that lineage-specific repression/silencing of these loci may result from localization to PCH, consistent with heterochromatic silencing. More intriguingly, we have found that the *Cd4* and *Cd8* loci dynamically associate during T cell development (Collins et al. 2011). The loci are closely associated in *cis* at the DP stage, separate slightly immediately after positive selection (CD4$^+$CD8^{lo} stage), and then are more closely associated in CD8$^+$ than in CD4$^+$ T cells. Association between the two loci is regulated by *cis*-acting elements in each locus (the E8$_I$ and E8$_{II}$ *Cd8* enhancers and *Cd4* silencer promote association in DPs and CD8$^+$ T cells), as well as the transcription factors that govern lineage choice (Runx proteins promote associations in DPs and CD8$^+$ T cells, while ThPOK antagonizes these associations in CD4$^+$ T cells). Intriguingly, this phenomenon is evolutionarily conserved, as the *CD4* and *CD8* loci associate closely in human CD8$^+$, but not CD4$^+$, T cells. It is tempting to speculate that long-distance association of the *Cd4* and *Cd8* loci in DP cells allows for co-receptor expression to be precisely and oppositely regulated during lineage commitment to the helper and cytotoxic lineages. This is indeed reminiscent of observations that the Th1 and Th2 cytokine loci interact in naïve T cells and separate upon polarization, and that the deletion of a DHS site in the Th2 locus delays activation of the Th1 locus during Th1 cell differentiation (Lee et al. 2005). Coordinated regulation of both cytokine and co-receptor loci may be facilitated by the close association of these loci in naïve helper T cells and immature DP cells, respectively.

3 Concluding Remarks

Study of co-receptor gene expression during T cell development has yielded insights into transcriptional regulatory mechanisms in general. While the expression of *Cd4* and *Cd8* is tightly coordinated throughout development, the molecular mechanisms that govern their simultaneous repression (DN stage), simultaneous expression (DP stage), and mutually exclusive expression (single positive stage) are quite different. At the DN stage *Cd4* is actively repressed by Runx1, while *Cd8* appears epigenetically repressed in a heterochromatin- and DNA methylation-dependent fashion. In DP cells, silencer function is abrogated to allow *Cd4* expression, while multiple enhancers are activated to overcome epigenetic silencing of *Cd8*. At this stage, expression of either gene is reversible. Finally, in helper T cells, transcription of *Cd4* is epigenetically maintained and *Cd8* expression is extinguished through unknown mechanisms, while in cytotoxic T cells *Cd4* is epigenetically silenced in a Runx3-dependent manner and *Cd8* transcription is presumably actively maintained through enhancer function. Thus to achieve two basic transcriptional outputs—on or off—for *Cd4* and *Cd8*, developing T cells appear to use multiple different mechanisms, some epigenetic and some not. This is reminiscent of chromatin profiling in Drosophila, which identified different "colors" or flavors of chromatin, defined by the occupancy of unique sets of chromatin-modifying complexes, chromatin readers, transcriptional regulators and transcription factors (Filion et al. 2010). Indeed these different chromatin colors were associated with different levels of transcription, as well as with genes belonging to different functional categories. For example, there were two types of active chromatin, one encompassing genes with broad expression patterns, and another enriched in genes linked to specific tissues. Detailed analysis of the protein structure and chromatin modifications at the *Cd4* and *Cd8* loci may reveal similar distinct types of chromatin flavors linked to active and repressed states (i.e. the *Cd4* locus in helper vs. cytotoxic cells), and instability or heritability of each of those states at different developmental stages (i.e. the *Cd4* locus in DP vs. helper T cells). Clearly, further characterization of these loci will yield critical insights into the epigenetic regulation of lineage-specific transcriptional programs.

References

Adlam M, Duncan DD, Ng DK, Siu G (1997) Positive selection induces CD4 promoter and enhancer function. Int Immunol 9:877–887

Adlam M, Siu G (2003) Hierarchical interactions control CD4 gene expression during thymocyte development. Immunity 18:173–184

Agalioti T, Lomvardas S, Parekh B, Yie J, Maniatis T, Thanos D (2000) Ordered recruitment of chromatin modifying and general transcription factors to the IFN-beta promoter. Cell 103:667–678

Allfrey VG (1966) Structural modifications of histones and their possible role in the regulation of ribonucleic acid synthesis. Proc Can Cancer Conf 6:313-335

Bachman KE, Park BH, Rhee I, Rajagopalan H, Herman JG, Baylin SB, Kinzler KW, Vogelstein B (2003) Histone modifications and silencing prior to DNA methylation of a tumor suppressor gene. Cancer Cell 3:89–95

Barski A, Cuddapah S, Cui K, Roh TY, Schones DE, Wang Z, Wei G, Chepelev I, Zhao K (2007) High-resolution profiling of histone methylations in the human genome. Cell 129:823–837

Bartke T, Vermeulen M, Xhemalce B, Robson SC, Mann M, Kouzarides T (2010) Nucleosome-interacting proteins regulated by DNA and histone methylation. Cell 143:470–484

The Epigenetic Landscape of Lineage Choice

Bilic I, Koesters C, Unger B, Sekimata M, Hertweck A, Maschek R, Wilson CB, Ellmeier W (2006) Negative regulation of CD8 expression via Cd8 enhancer-mediated recruitment of the zinc finger protein MAZR. Nat Immunol 7:392–400

Blum MD, Wong GT, Higgins KM, Sunshine MJ, Lacy E (1993) Reconstitution of the subclass-specific expression of CD4 in thymocytes and peripheral T cells of transgenic mice: identification of a human CD4 enhancer. J Exp Med 177:1343–1358

Brickner DG, Cajigas I, Fondufe-Mittendorf Y, Ahmed S, Lee PC, Widom J, Brickner JH (2007) H2A.Z-mediated localization of genes at the nuclear periphery confers epigenetic memory of previous transcriptional state. PLoS Biol 5:e81

Brown CJ, Willard HF (1994) The human X-inactivation centre is not required for maintenance of X-chromosome inactivation. Nature 368:154–156

Bruniquel D, Borie N, Hannier S, Triebel F (1998) Regulation of expression of the human lymphocyte activation gene-3 (LAG-3) molecule, a ligand for MHC class II. Immunogenetics 48:116–124

Bruniquel D, Schwartz RH (2003) Selective, stable demethylation of the interleukin-2 gene enhances transcription by an active process. Nat Immunol 4:235–240

Bruno L, Mazzarella L, Hoogenkamp M, Hertweck A, Cobb BS, Sauer S, Hadjur S, Leleu M, Naoe Y, Telfer JC et al (2009) Runx proteins regulate Foxp3 expression. J Exp Med 206:2329–2337

Campos EI, Reinberg D (2009) Histones: annotating chromatin. Annu Rev Genet 43:559–599

Carbone AM, Marrack P, Kappler JW (1988) Demethylated CD8 gene in CD4$^+$ T cells suggests that CD4$^+$ cells develop from CD8$^+$ precursors. Science 242:1174–1176

Chi TH, Wan M, Lee PP, Akashi K, Metzger D, Chambon P, Wilson CB, Crabtree GR (2003) Sequential roles of Brg, the ATPase subunit of BAF chromatin remodeling complexes, in thymocyte development. Immunity 19:169–182

Chi TH, Wan M, Zhao K, Taniuchi I, Chen L, Littman DR, Crabtree GR (2002) Reciprocal regulation of CD4/CD8 expression by SWI/SNF-like BAF complexes. Nature 418:195–199

Chong MM, Simpson N, Ciofani M, Chen G, Collins A, Littman DR (2010) Epigenetic propagation of CD4 expression is established by the Cd4 proximal enhancer in helper T cells. Genes Dev 24:659–669

Ciofani M, Zuniga-Pflucker JC (2010) Determining gammadelta versus alphass T cell development. Nat Rev Immunol 10:657–663

Collins A, Hewitt SL, Chaumeil J, Sellars M, Micsinai M, Allinne J, Parisi F, Nora EP, Bolland DJ, Corcoran AE et al (2011) RUNX transcription factor-mediated association of Cd4 and Cd8 enables coordinate gene regulation. Immunity 34:303–314

Csankovszki G, Panning B, Bates B, Pehrson JR, Jaenisch R (1999) Conditional deletion of Xist disrupts histone macroH2A localization but not maintenance of X inactivation. Nat Genet 22:323–324

Delaire S, Huang YH, Chan SW, Robey EA (2004) Dynamic repositioning of CD4 and CD8 genes during T cell development. J Exp Med 200:1427–1435

Devine L, Kieffer LJ, Aitken V, Kavathas PB (2000) Human CD8 beta, but not mouse CD8 beta, can be expressed in the absence of CD8 alpha as a beta beta homodimer. J Immunol 164: 833–838

Egawa T, Littman DR (2008) ThPOK acts late in specification of the helper T cell lineage and suppresses Runx-mediated commitment to the cytotoxic T cell lineage. Nat Immunol 9: 1131–1139

Egawa T, Littman DR (2011) The transcription factor AP4 modulates reversible and epigenetic silencing of the *Cd4* gene. Proc Natl Acad Sci USA 108:14873–14878

Egawa T, Tillman RE, Naoe Y, Taniuchi I, Littman DR (2007) The role of the Runx transcription factors in thymocyte differentiation and in homeostasis of naive T cells. J Exp Med 204: 1945–1957

Ellmeier W, Sunshine MJ, Losos K, Hatam F, Littman DR (1997) An enhancer that directs lineage-specific expression of CD8 in positively selected thymocytes and mature T cells. Immunity 7:537–547

Ellmeier W, Sunshine MJ, Losos K, Littman DR (1998) Multiple developmental stage-specific enhancers regulate CD8 expression in developing thymocytes and in thymus-independent T cells. Immunity 9:485–496

Ellmeier W, Sunshine MJ, Maschek R, Littman DR (2002) Combined deletion of CD8 locus cis-regulatory elements affects initiation but not maintenance of CD8 expression. Immunity 16:623–634

Feik N, Bilic I, Tinhofer J, Unger B, Littman DR, Ellmeier W (2005) Functional and molecular analysis of the double-positive stage-specific CD8 enhancer E8III during thymocyte development. J Immunol 174:1513–1524

Ficz G, Branco MR, Seisenberger S, Santos F, Krueger F, Hore TA, Marques CJ, Andrews S, Reik W (2011) Dynamic regulation of 5-hydroxymethylcytosine in mouse ES cells and during differentiation. Nature 473:398–402

Filion GJ, van Bemmel JG, Braunschweig U, Talhout W, Kind J, Ward LD, Brugman W, de Castro IJ, Kerkhoven RM, Bussemaker HJ et al (2010) Systematic protein location mapping reveals five principal chromatin types in Drosophila cells. Cell 143:212–224

Fischle W, Tseng BS, Dormann HL, Ueberheide BM, Garcia BA, Shabanowitz J, Hunt DF, Funabiki H, Allis CD (2005) Regulation of HP1-chromatin binding by histone H3 methylation and phosphorylation. Nature 438:1116–1122

Fodor BD, Shukeir N, Reuter G, Jenuwein T (2010) Mammalian Su(var) genes in chromatin control. Annu Rev Cell Dev Biol 26:471–501

Foster SL, Hargreaves DC, Medzhitov R (2007) Gene-specific control of inflammation by TLR-induced chromatin modifications. Nature 447:972–978

Frankel N, Davis GK, Vargas D, Wang S, Payre F, Stern DL (2010) Phenotypic robustness conferred by apparently redundant transcriptional enhancers. Nature 466:490–493

Fuks F, Burgers WA, Brehm A, Hughes-Davies L, Kouzarides T (2000) DNA methyltransferase Dnmt1 associates with histone deacetylase activity. Nat Genet 24:88–91

Fung-Leung WP, Schilham MW, Rahemtulla A, Kundig TM, Vollenweider M, Potter J, van Ewijk W, Mak TW (1991) CD8 is needed for development of cytotoxic T cells but not helper T cells. Cell 65:443–449

Garefalaki A, Coles M, Hirschberg S, Mavria G, Norton T, Hostert A, Kioussis D (2002) Variegated expression of CD8 alpha resulting from in situ deletion of regulatory sequences. Immunity 16:635–647

Gialitakis M, Arampatzi P, Makatounakis T, Papamatheakis J (2010) Gamma interferon-dependent transcriptional memory via relocalization of a gene locus to PML nuclear bodies. Mol Cell Biol 30:2046–2056

Hamerman JA, Page ST, Pullen AM (1997) Distinct methylation states of the CD8 beta gene in peripheral T cells and intraepithelial lymphocytes. J Immunol 159:1240–1246

Hanna Z, Simard C, Laperriere A, Jolicoeur P (1994) Specific expression of the human CD4 gene in mature CD4$^+$ CD8$^-$ and immature CD4$^+$ CD8$^+$ T cells and in macrophages of transgenic mice. Mol Cell Biol 14:1084–1094

Hansen KH, Bracken AP, Pasini D, Dietrich N, Gehani SS, Monrad A, Rappsilber J, Lerdrup M, Helin K (2008) A model for transmission of the H3K27me3 epigenetic mark. Nat Cell Biol 10:1291–1300

He HH, Meyer CA, Shin H, Bailey ST, Wei G, Wang Q, Zhang Y, Xu K, Ni, M, Lupien M et al (2010) Nucleosome dynamics define transcriptional enhancers. Nat Genet 42:343–347

He X, Dave VP, Zhang Y, Hua X, Nicolas E, Xu W, Roe BA, Kappes DJ (2005) The zinc finger transcription factor Th-POK regulates CD4 versus CD8 T-cell lineage commitment. Nature 433:826–833

Heintzman ND, Stuart RK, Hon G, Fu Y, Ching CW, Hawkins RD, Barrera LO, Van Calcar S, Qu C, Ching KA et al (2007) Distinct and predictive chromatin signatures of transcriptional promoters and enhancers in the human genome. Nat Genet 39:311–318

Hernandez-Hoyos G, Anderson MK, Wang C, Rothenberg EV, Alberola-Ila J (2003) GATA-3 expression is controlled by TCR signals and regulates CD4/CD8 differentiation. Immunity 19:83–94

Hostert A, Garefalaki A, Mavria G, Tolaini M, Roderick K, Norton T, Mee PJ, Tybulewicz VL, Coles M, Kioussis D (1998) Hierarchical interactions of control elements determine CD8alpha gene expression in subsets of thymocytes and peripheral T cells. Immunity 9:497–508

Hostert A, Tolaini M, Festenstein R, McNeill L, Malissen B, Williams O, Zamoyska R, Kioussis D (1997a) A CD8 genomic fragment that directs subset-specific expression of CD8 in transgenic mice. J Immunol 158:4270–4281

Hostert A, Tolaini M, Roderick K, Harker N, Norton T, Kioussis D (1997b) A region in the CD8 gene locus that directs expression to the mature CD8 T cell subset in transgenic mice. Immunity 7:525–536

Jeon Y, Lee JT (2011) YY1 Tethers Xist RNA to the Inactive X Nucleation Center. Cell 146:119–133

Ji H, Ehrlich LI, Seita J, Murakami P, Doi A, Lindau P, Lee H, Aryee MJ, Irizarry RA, Kim K et al (2010) Comprehensive methylome map of lineage commitment from haematopoietic progenitors. Nature 467:338–342

Jiang H, Peterlin BM (2008) Differential chromatin looping regulates CD4 expression in immature thymocytes. Mol Cell Biol 28:907–912

Jiang H, Zhang F, Kurosu T, Peterlin BM (2005) Runx1 binds positive transcription elongation factor b and represses transcriptional elongation by RNA polymerase II: possible mechanism of CD4 silencing. Mol Cell Biol 25:10675–10683

Jones ME, Zhuang Y (2007) Acquisition of a functional T cell receptor during T lymphocyte development is enforced by HEB and E2A transcription factors. Immunity 27:860–870

Kieffer LJ, Bennett JA, Cunningham AC, Gladue RP, McNeish J, Kavathas PB, Hanke JH (1996) Human CD8 alpha expression in NK cells but not cytotoxic T cells of transgenic mice. Int Immunol 8:1617–1626

Kieffer LJ, Yan L, Hanke JH, Kavathas PB (1997) Appropriate developmental expression of human CD8 beta in transgenic mice. J Immunol 159:4907–4912

Killeen N, Sawada S, Littman DR (1993) Regulated expression of human CD4 rescues helper T cell development in mice lacking expression of endogenous CD4. EMBO J 12:1547–1553

Lal G, Zhang N, van der Touw W, Ding Y, Ju W, Bottinger EP, Reid SP, Levy DE, Bromberg JS (2009) Epigenetic regulation of Foxp3 expression in regulatory T cells by DNA methylation. J Immunol 182:259–273

Lee CK, Shibata Y, Rao B, Strahl BD, Lieb JD (2004) Evidence for nucleosome depletion at active regulatory regions genome-wide. Nat Genet 36:900–905

Lee GR, Spilianakis CG, Flavell RA (2005) Hypersensitive site 7 of the TH2 locus control region is essential for expressing TH2 cytokine genes and for long-range intrachromosomal interactions. Nat Immunol 6:42–48

Lee JS, Smith E, Shilatifard A (2010) The language of histone crosstalk. Cell 142:682–685

Lee PP, Fitzpatrick DR, Beard C, Jessup HK, Lehar S, Makar KW, Perez-Melgosa M, Sweetser MT, Schlissel MS, Nguyen S et al (2001) A critical role for Dnmt1 and DNA methylation in T cell development, function, and survival. Immunity 15:763-774

Leung RK, Thomson K, Gallimore A, Jones E, Van den Broek M, Sierro S, Alsheikhly AR, McMichael A, Rahemtulla A (2001) Deletion of the CD4 silencer element supports a stochastic mechanism of thymocyte lineage commitment. Nat Immunol 2:1167–1173

Levanon D, Goldstein RE, Bernstein Y, Tang H, Goldenberg D, Stifani S, Paroush Z, Groner Y (1998) Transcriptional repression by AML1 and LEF-1 is mediated by the TLE/Groucho corepressors. Proc Natl Acad Sci USA 95:11590–11595

Lomvardas S, Thanos D (2001) Nucleosome sliding via TBP DNA binding in vivo. Cell 106:685–696

Manjunath N, Shankar P, Stockton B, Dubey PD, Lieberman J, von Andrian UH (1999) A transgenic mouse model to analyze CD8(+) effector T cell differentiation in vivo. Proc Natl Acad Sci U S A 96:13932–13937

Margueron R, Justin N, Ohno K, Sharpe ML, Son J, Drury WJ, 3rd, Voigt P, Martin SR, Taylor WR, De Marco V et al (2009) Role of the polycomb protein EED in the propagation of repressive histone marks. Nature 461:762–767

Mateescu B, Bourachot B, Rachez C, Ogryzko V, Muchardt C (2008) Regulation of an inducible promoter by an HP1beta-HP1gamma switch. EMBO Rep 9:267–272

Merkenschlager M, Amoils S, Roldan E, Rahemtulla A, O'Connor E, Fisher AG, Brown KE (2004) Centromeric repositioning of coreceptor loci predicts their stable silencing and the CD4/CD8 lineage choice. J Exp Med 200:1437–1444

Murayama A, Sakura K, Nakama M, Yasuzawa-Tanaka K, Fujita E, Tateishi Y, Wang Y, Ushijima T, Baba T, Shibuya K et al (2006) A specific CpG site demethylation in the human interleukin 2 gene promoter is an epigenetic memory. EMBO J 25:1081–1092

Muroi S, Naoe Y, Miyamoto C, Akiyama K, Ikawa T, Masuda K, Kawamoto H, Taniuchi I (2008) Cascading suppression of transcriptional silencers by ThPOK seals helper T cell fate. Nat Immunol 9:1113–1121

Naito T, Gomez-Del Arco P, Williams CJ, Georgopoulos K (2007) Antagonistic interactions between Ikaros and the chromatin remodeler Mi-2beta determine silencer activity and Cd4 gene expression. Immunity 27:723–734

Naito T, Taniuchi I (2010) The network of transcription factors that underlie the CD4 versus CD8 lineage decision. Int Immunol 22:791–796

Ng HH, Robert F, Young RA, Struhl K (2003) Targeted recruitment of Set1 histone methylase by elongating Pol II provides a localized mark and memory of recent transcriptional activity. Mol Cell 11:709–719

Ng RK, Gurdon JB (2008) Epigenetic memory of an active gene state depends on histone H3.3 incorporation into chromatin in the absence of transcription. Nat Cell Biol 10:102–109

O'Neill MJ (2005) The influence of non-coding RNAs on allele-specific gene expression in mammals. Hum Mol Genet 14 Spec No 1, R113–120

Olins AL, Olins DE (1974) Spheroid chromatin units (v bodies). Science 183:330–332

Orphanides G, Reinberg D (2000) RNA polymerase II elongation through chromatin. Nature 407:471–475

Osborne CS, Chakalova L, Brown KE, Carter D, Horton A, Debrand E, Goyenechea B, Mitchell JA, Lopes S, Reik W et al (2004) Active genes dynamically colocalize to shared sites of ongoing transcription. Nat Genet 36:1065–1071

Ozsolak F, Song JS, Liu XS, Fisher DE (2007) High-throughput mapping of the chromatin structure of human promoters. Nat Biotechnol 25:244–248

Pai SY, Truitt ML, Ting CN, Leiden JM, Glimcher LH, Ho IC (2003) Critical roles for transcription factor GATA-3 in thymocyte development. Immunity 19:863–875

Petesch SJ, Lis JT (2008) Rapid, transcription-independent loss of nucleosomes over a large chromatin domain at Hsp70 loci. Cell 134:74–84

Polansky JK, Kretschmer K, Freyer J, Floess S, Garbe A, Baron U, Olek S, Hamann A, von Boehmer H, Huehn J (2008) DNA methylation controls Foxp3 gene expression. Eur J Immunol 38:1654–1663

Reed-Inderbitzin E, Moreno-Miralles I, Vanden-Eynden SK, Xie J, Lutterbach B, Durst-Goodwin KL, Luce KS, Irvin BJ, Cleary ML, Brandt SJ et al (2006) RUNX1 associates with histone deacetylases and SUV39H1 to repress transcription. Oncogene 25:5777–5786

Roh TY, Wei G, Farrell CM, Zhao K (2007) Genome-wide prediction of conserved and nonconserved enhancers by histone acetylation patterns. Genome Res 17:74–81

Rothenberg EV, Moore JE, Yui MA (2008) Launching the T-cell-lineage developmental programme. Nat Rev Immunol 8:9–21

Sado T, Fenner MH, Tan SS, Tam P, Shioda T, Li E (2000) X inactivation in the mouse embryo deficient for Dnmt1: distinct effect of hypomethylation on imprinted and random X inactivation. Dev Biol 225:294–303

Sakaguchi S, Hombauer M, Bilic I, Naoe Y, Schebesta A, Taniuchi I, Ellmeier W (2010) The zinc-finger protein MAZR is part of the transcription factor network that controls the CD4 versus CD8 lineage fate of double-positive thymocytes. Nat Immunol 11:442–448

Sands JF, Nikolic-Zugic J (1992) T cell-specific protein-DNA interactions occurring at the CD4 locus: identification of possible transcriptional control elements of the murine CD4 gene. Int Immunol 4:1183–1194

Sarafova SD, Erman B, Yu Q, Van Laethem F, Guinter T, Sharrow SO, Feigenbaum L, Wildt KF, Ellmeier W, Singer A (2005) Modulation of coreceptor transcription during positive selection dictates lineage fate independently of TCR/coreceptor specificity. Immunity 23:75–87

Sato T, Ohno S, Hayashi T, Sato C, Kohu K, Satake M, Habu S (2005) Dual functions of Runx proteins for reactivating CD8 and silencing CD4 at the commitment process into CD8 thymocytes. Immunity 22:317–328

Sawada S, Littman DR (1991) Identification and characterization of a T-cell-specific enhancer adjacent to the murine CD4 gene. Mol Cell Biol 11:5506–5515

Sawada S, Littman DR (1993) A heterodimer of HEB and an E12-related protein interacts with the CD4 enhancer and regulates its activity in T-cell lines. Mol Cell Biol 13:5620–5628

Sawada S, Scarborough JD, Killeen N, Littman DR (1994) A lineage-specific transcriptional silencer regulates CD4 gene expression during T lymphocyte development. Cell 77:917–929

Singer A, Adoro S, Park JH (2008) Lineage fate and intense debate: myths, models and mechanisms of CD4- versus CD8-lineage choice. Nat Rev Immunol 8:788–801

Siu G, Wurster AL, Duncan DD, Soliman TM, Hedrick SM (1994) A transcriptional silencer controls the developmental expression of the CD4 gene. EMBO J 13:3570–3579

Spilianakis CG, Lalioti MD, Town T, Lee GR, Flavell RA (2005) Interchromosomal associations between alternatively expressed loci. Nature 435:637–645

Starr TK, Jameson SC, Hogquist KA (2003) Positive and negative selection of T cells. Annu Rev Immunol 21:139–176

Sun G, Liu X, Mercado P, Jenkinson SR, Kypriotou M, Feigenbaum L, Galera P, Bosselut R (2005) The zinc finger protein cKrox directs CD4 lineage differentiation during intrathymic T cell positive selection. Nat Immunol 6:373–381

Tahiliani M, Koh KP, Shen Y, Pastor WA, Bandukwala H, Brudno Y, Agarwal S, Iyer LM, Liu DR, Aravind L et al (2009) Conversion of 5-methylcytosine to 5-hydroxymethylcytosine in mammalian DNA by MLL partner TET1. Science 324:930–935

Taniuchi I, Ellmeier W, Littman DR (2004) The CD4/CD8 lineage choice: new insights into epigenetic regulation during T cell development. Adv Immunol 83:55–89

Taniuchi I, Osato M, Egawa T, Sunshine MJ, Bae SC, Komori T, Ito Y, Littman DR (2002a) Differential requirements for Runx proteins in CD4 repression and epigenetic silencing during T lymphocyte development. Cell 111:621–633

Taniuchi I, Sunshine MJ, Festenstein R, Littman DR (2002b) Evidence for distinct CD4 silencer functions at different stages of thymocyte differentiation. Mol Cell 10:1083–1096

Telfer JC, Hedblom EE, Anderson MK, Laurent MN, Rothenberg EV (2004) Localization of the domains in Runx transcription factors required for the repression of CD4 in thymocytes. J Immunol 172:4359–4370

Wan M, Zhang J, Lai D, Jani A, Prestone-Hurlburt P, Zhao L, Ramachandran A, Schnitzler GR, and Chi T (2009) Molecular basis of CD4 repression by the Swi/Snf-like BAF chromatin remodeling complex. Eur J Immunol 39:580–588

Wang L, Wildt KF, Castro E, Xiong Y, Feigenbaum L, Tessarollo L, Bosselut R (2008a) The zinc finger transcription factor Zbtb7b represses CD8-lineage gene expression in peripheral CD4$^+$ T cells. Immunity 29:876–887

Wang L, Wildt KF, Zhu J, Zhang X, Feigenbaum L, Tessarollo L, Paul WE, Fowlkes BJ, Bosselut R (2008b) Distinct functions for the transcription factors GATA-3 and ThPOK during intrathymic differentiation of CD4(+) T cells. Nat Immunol 9:1122–1130

Williams LM, Rudensky AY (2007) Maintenance of the Foxp3-dependent developmental program in mature regulatory T cells requires continued expression of Foxp3. Nat Immunol 8:277–284

Woodcock CL, Ghosh RP (2010) Chromatin higher-order structure and dynamics. Cold Spring Harb Perspect Biol 2:a000596

Wu C, Wong YC, Elgin SC (1979) The chromatin structure of specific genes: II. Disruption of chromatin structure during gene activity. Cell 16:807–814

Wu H, Coskun V, Tao J, Xie W, Ge W, Yoshikawa K, Li E, Zhang Y, Sun YE (2010) Dnmt3a-dependent nonpromoter DNA methylation facilitates transcription of neurogenic genes. Science 329:444–448

Wurster AL, Siu G, Leiden JM, Hedrick SM (1994) Elf-1 binds to a critical element in a second CD4 enhancer. Mol Cell Biol 14:6452–6463

Wutz A, Jaenisch R (2000) A shift from reversible to irreversible X inactivation is triggered during ES cell differentiation. Mol Cell 5:695–705

Yarmus M, Woolf E, Bernstein Y, Fainaru O, Negreanu V, Levanon D, Groner Y (2006) Groucho/transducin-like enhancer-of-split (TLE)-dependent and -independent transcriptional regulation by Runx3. Proc Natl Acad Sci U S A 103:7384–7389

Yu M, Wan M, Zhang J, Wu J, Khatri R, Chi T (2008) Nucleoprotein structure of the CD4 locus: implications for the mechanisms underlying CD4 regulation during T cell development. Proc Natl Acad Sci U S A 105:3873–3878

Zhang XL, Seong R, Piracha R, Larijani M, Heeney M, Parnes JR, Chamberlain JW (1998) Distinct stage-specific cis-active transcriptional mechanisms control expression of T cell coreceptor CD8 alpha at double- and single-positive stages of thymic development. J Immunol 161:2254–2266

Zhang XL, Zhao S, Borenstein SH, Liu Y, Jayabalasingham B, Chamberlain JW (2001) CD8 expression up to the double-positive CD3(low/intermediate) stage of thymic differentiation is sufficient for development of peripheral functional cytotoxic T lymphocytes. J Exp Med 194:685–693

Zheng Y, Josefowicz S, Chaudhry A, Peng XP, Forbush K, Rudensky AY (2010) Role of conserved non-coding DNA elements in the Foxp3 gene in regulatory T-cell fate. Nature 463:808–812

Zou YR, Sunshine MJ, Taniuchi I, Hatam F, Killeen N, Littman DR (2001) Epigenetic silencing of CD4 in T cells committed to the cytotoxic lineage. Nat Genet 29:332–336

Index

30 nm fiber, 168
3' regulatory region (3'RR), 44
3C, 44
5'DFL, 44

A
Accessibility hypothesis, 42
Accessibility, 127
Akt, 17
Allelic exclusion, 52
Alternative modes of repression, 134
Antisense, 71
AP4, 173

B
βselection, 167
BCR, 17, 29
Beads-on-a-string, 168
Bromodomains, 169

C
C/EBP family factors, 127
CD19, 29
CD4, 166–168, 171, 176–177, 181–182
CD79a, 18, 20, 25, 30, 33
CD8, 165–168, 176, 178–179, 181–182
Cebpa, 27
Chip-on-chip, 28
ChIP-Seq, 2, 29
Chromatin, 42, 91, 93, 95, 97–99, 101–110
Chromodomains, 169
Chromosome looping, 55

Cis regulatory elements, 2
Cistromes, 2
CLP, 23, 25, 34
Combinatorial gene control, 120
Cooperative DNA binding, 4
CpG, 33
CpG methylation, 131
C/EBP family factors, 127
CTCF, 43

D
DFL16.1, 40
Distal, 48
DNA demethylation , 33
DNA hypomethylation, 178
DNA methylation, 19, 31, 169, 175, 180–181
DNase hypersensitivity, 123
Dnmt1, 180–181
DQ52, 40
DSP, 40

E
E2A, 18–19, 23, 29–31, 33–34, 126, 173
EBF1, 17–20, 24–25, 28–30, 33–34, 127
EBF3, 20
Eμ, 43
Enhancer, 9, 42
Epigenetic marks, 170
Epigenetic regulation, 30, 170–171
Epigenetic remodeling, 19
Epigenetic, 42

190 Index

E (*cont.*)
Euchromatin, 168
Ezh2, 54

F
Feed back loop, 25
FOXO1, 19, 25

G
GATA-3, 123, 167
Gfi-1, 23
Groucho/TLE, 176

H
H2A.Z, 178
H3 acetylation, 33
H3.3, 178
H3Ac, 169
H3K27, 168, 170, 176
H3K27me3, 128
H3K4me1, 30
H3K4me1$^+$, 11
H3K4me1^{hi}, 9
H3K4me2, 30
H3K4me3^{lo}, 9
H3K7, 168
H3K79, 169
H3S10, 169
H4Ac, 169
H4K20, 169
Haploinsufficiency, 27
HEB, 173
Hes1, 29
Heterochromatic silencing, 181
Heterochromatin, 43, 168, 175
Histone acetyl transferase, 33, 169
Histone deacetyl transferase, 169
Histone variants, 178
Hit-and-Run Gene Regulation, 121
HLHLH domain, 19, 33
HP1, 169, 175
HSC, 24

I
Id2, 27
Id3, 27
Ifng, 122, 123
Ikaros, 23, 54
Il2, 122
Il4, 122, 123

IL-4, 123
IL7, 25
Immunoglobulin heavy chain, 65
Instructive, 167
Intergenic, 71

L
Lineage choice, 167
Lineage priming, 23–24
Lineage restriction, 132
Lineage-specific gene expression, 119
LMPP, 23
Locus compaction, 55
Lternative modes of repression, 134

M
MAZR, 168, 179
Methylation, 168
Methylation of H3K4, 125
MHCI, 167
MHCII, 167
Mi-2/NuRD, 31, 33–34
Mi-2β, 33
MLL, 10
mTOR, 17
Multilineage priming, 130
Myb, 168

N
NK cell development, 28
Non-coding RNA transcription, 65
Notch, 130
Nuclear bodies, 170
Nuclear organization, 65
Nucleosome density, 33
NuRD, 173

P
Pax5, 18, 24–25, 29, 32–33, 53
Peri-centormeric heterochromatin, 181
PHD fingers, 169
Phosphorylation, 168
PI3K, 17, 29–30
Pioneer Factors, 6, 31
PML, 170
Polycomb Repression Complex, 128
Positive feedback, 122
Pou2af1, 25
PQ52, 43
PRC2 methyltransferase, 170

Index

Pre-BCR, 17
Promoters, 42
Proximal, 47
pTEFb, 173
PU.1, 23, 126

R
RAG, 44
Recombination center, 44
Recombination, 40
Redundant enhancers, 179
Regulation, 121
Repertoire, 52
Repression, 129, 131
RSSs, 41
Runx, 19, 25, 31, 33, 129
Runx1, 173, 176
Runx3, 167, 173, 182

S
Sequential epigenetic analysis, 134
Silencing, 132
Site selectivity, 127
STAT signaling, 25
Stat4, 123
STAT5, 53, 168
Stat6, 123
Stem and progenitor cell, 132
Stem/progenitor regulatory genes, 132
Sterile transcripts, 44
Stochastic, 167
Sumoylation, 168
SUV39H1 H3K9 methyltransferase, 176
SWI/SNF, 19, 33, 170, 173, 175, 179

T
T cell receptor genes, 92
T lineage commitment, 131

T-bet (Tbx21), 123
TCR $\gamma\delta$, 166
TCR $\alpha\beta$, 166
Ternary Complex, 3, 4, 33
Tet proteins, 169
Th1, 181
Th2, 122, 181
ThPOK, 167
Thymocytes, 93–95, 97, 99, 101–104,
        107–110
TIG/IPT, 22, 33
Tox, 168
Transcription Factor Concentrations, 7
Transcription factories, 170
Transcription factors, 120
Transcription, 96–102, 104, 105, 109, 110
Transcriptional network, 167

U
Ubiquitination, 168

V
V(D)J recombination, 65, 92–95, 106, 108
Variegated expression, 175, 180

W
WD40 repeats, 169

Y
YY1, 53

Z
Zinc knuckle, 20, 22, 33